Ambrose Loomis Ranney

The Topographical Relations of the Female Pelvic Organs

Ambrose Loomis Ranney

The Topographical Relations of the Female Pelvic Organs

ISBN/EAN: 9783743414860

Manufactured in Europe, USA, Canada, Australia, Japa

Cover: Foto ©berggeist007 / pixelio.de

Manufactured and distributed by brebook publishing software
(www.brebook.com)

Ambrose Loomis Ranney

The Topographical Relations of the Female Pelvic Organs

THE

TOPOGRAPHICAL RELATIONS

OF THE

FEMALE PELVIC ORGANS

BY

AMBROSE L. RANNEY, A.M., M.D.,

ADJUNCT PROFESSOR OF ANATOMY IN THE MEDICAL DEPARTMENT OF THE
UNIVERSITY OF THE CITY OF NEW YORK; AUTHOR OF "THE
APPLIED ANATOMY OF THE NERVOUS SYSTEM," "A PRAC-
TICAL TREATISE ON SURGICAL DIAGNOSIS," ETC.

WITH 22 WOODCUTS.

NEW YORK:
WM. WOOD & CO., PUBLISHERS, 56 & 58 LAFAYETTE PLACE.
1883.

THE
TOPOGRAPHICAL RELATIONS
OF THE
FEMALE PELVIC ORGANS.

THE art of gynecology, in contrast with some other departments of medical science, is one of comparatively recent date. With the growth of this branch and the development of certain surgical procedures, which seem destined to surpass in their results even the most sanguine expectations of those who first devised them, has arisen a feeling of desire on the part of most surgeons in this special field for an accurate and complete description of the parts which come more particularly under their observation. There is, I believe, a realizing sense among the profession that anatomy has not kept pace in its growth with the other departments of gynecology. It is true that the microscope has enabled some investigators to add much to our knowledge of the minute structure of various tissues which compose the pelvic organs; and the description of the general contour and construction of these organs has been worked out in some detail by most of our leading anatomists. Nevertheless it is a fact which cannot be disputed by any one familiar with the literature of the subject, that the *topographical* descriptions of parts most essential to a practical gynæcologist are as yet largely a matter of dispute even among authors of acknowledged reputation.

In this special department of surgery, anatomy has a most important bearing upon successful treatment. As we pass inward, in our investigation, from the tissues which help to form the pelvic floor toward those which enter into the formation of the vagina, uterus, Fallopian tubes, ovaries, urethral canal, bladder, and rectum, not to speak of the various fasciæ, the pelvic cellular tissue and peritoneum, the ligamentous structures connected with the organs mentioned, and the blood-vessels, lymphatics, and nerves, it is impossible to single out one which is not to-day an open field for scientific discovery and which cannot suggest to the progressive student many points that are of practical value.

I think it may be claimed, without saying more than facts will apparently justify, that the anatomy of the pelvic organs and the various structures in intimate relation with them is more difficult than of that of any other topographical region of the body. The physiological functions which these parts are called upon to perform—requiring as they do an inherent mobility without a liability to displacement, a capacity for marked alteration in their relations to each other during the pregnant state without endangering their subsequent return to the normal standard, a strength and elasticity which shall enable certain parts to sustain the weight of super-incumbent organs without damage to the important parts in the neighborhood of the pelvic brim, an apparently open pelvic floor which shall resist all tendency to hernial protrusions of the organs supported—all these and many other evidences of the wisdom of the Creator must strike the most superficial student of anatomy with awe, which is heightened rather than decreased by careful research.

It is with a full sense of the responsibility of the task before me that I attempt to present to the profession a *résumé* of what I believe to be the true anatomy of the regions which properly pertain to the department of gynæcology. In addition to my own researches in this field, made some years since upon the cadaver, and lately in the autopsy rooms of hospitals, I shall bring to my aid all that has been done in the same field by those more competent than myself—endeavoring to do full justice to their opinions, but reserving the right allowed each author (which I trust will be used in an unprejudiced and impartial spirit), of careful discrimination between what seems to me to be established fact and unsupported theory.

The anatomy of the pelvic organs (to be thoroughly grasped by the comprehension in order that it may be applied to various diseased conditions) must be so presented as to faithfully portray not merely the tissues which form special parts, but also the relations which each bears to those in its immediate vicinity. Thus it becomes necessary to know not only where the viscera are situated, and their modifications in the state of health (since some are displaced at times by the distention of neighboring organs and by other physiological processes), but also the exact position of all the structures which are associated with each, either by simple contiguity or its physiological functions.

If we critically examine drawings, found in standard textbooks, which are supposed to portray accurately the parts which they are intended to represent, it must be apparent to all that some must be grossly inaccurate; since it is difficult to find any two which agree in all important points. The same criticism may be made, although perhaps to a less degree, as to representations given in some of the more recent and scientific monographs upon the subject. If we seek to explain these apparent discrepancies only on the ground of lack of care in the drawing [1] or inaccuracy in the dissections made, a serious injustice may, in many instances, be done to some authors whose skill as anatomists must be above question, and to artists whose fidelity to nature is well known. It will be my unpleasant duty to differ widely from some in my conception of the topography of the pelvis in its antero-posterior median section, and to criticise the drawings of others as grossly incorrect; I may be pardoned, therefore, for an attempt to explain what to my mind are the chief causes which conduce toward error in description and proper delineation of the pelvis and its contents.

---

[1] The inaccuracies which exist in some cuts and plates (often referred to by authors in commendatory terms) are so gross as to merit severe censure on this ground.

CAUSES OF ERROR AS TO THE ACCURACY OF EXISTING PLATES AND CUTS.—In the first place, I believe that too great reliance should not be placed upon the accuracy of any *dissection*, as a step toward determining the relations of these organs. The tissues of a cadaver have lost many of the essential elements which live tissues possess—chiefly their elasticity—and thus the weight alone of the super-incumbent structures and the formation of gases in the intestine may assist in creating a displacement of the organs which are being investigated, from their normal surroundings. Again, the fact that attitude has a disturbing influence upon these organs is well known; and no dissection of a body made in the recumbent posture is a proper guide for a drawing intended to portray the normal position of the viscera in the erect attitude. As an illustration of this fact, I have discovered that, in several of the most universally quoted works upon the subject, the pictures represent the woman as in the standing posture, and yet the pelvis is shown in a greatly distorted position in reference to the angle which it should normally make with the spinal column, while the sacrum and pubes are also incorrectly placed. No one believes to-day that the vagina is an open tube in the erect attitude of the woman, or in point of fact in any other posture, either during life or after death; and yet how often do we see it so depicted in cuts of supposed merit. I can conceive of no better way to determine the ramifications of a fascia, the attachments of a muscle, or the course of a nerve or blood-vessel, than by dissection; but, if plates (as usually made from carefully prepared dissections) are to be considered as faithful portraits, this plan of instructing the surgeon as to topographical relation seems to me comparatively worthless, in case the parts examined are extremely movable or liable to be disturbed from their normal relations to adjacent structures. Finally, it is absolutely impossible to expose some of the pelvic viscera without dividing structures which are essential to the maintenance of their normal relations to neighboring parts; hence dissections are in themselves a source of decided error in many instances, even if carefully prepared.

Of late, an attempt has been made by some investigators to eliminate these sources of error by freezing a corpse and then making sections which can be duplicated by photography. This is an unquestioned improvement upon the old method by dissection, but it is open to many of the objections previously made. The tissues have not only lost their elasticity by death, but they may possibly have been distorted by the action of intense cold. The recumbent posture was probably that in which the subject died—certainly that in which she must have, of necessity, been laid immediately after death—and we have no reason to believe that the organs of the pelvis will ever regain their normal position and relations to surrounding parts in a corpse, when the weight of the abdominal viscera and the action of gravity have probably displaced them. This will not be insured even after the body has been placed in the "knee-chest" position (as has been done by some of the later investigators)—probably after a prolonged dorsal decubitus. I do not totally decry this method of research, as I deem it of great value for some purposes; but I am loth to accept the sections so represented as an absolute exposition of the organs of this region in the state of health.[1]

Finally, I believe that much error in the future drawings of the antero-posterior median section of the pelvis is to be eliminated by a more extended series of investigation upon the living subject than gynecological literature as yet can produce. It is most certainly possible, by methods which have already been published in some detail,[2] to determine many of the essential measurements, upon which a drawing of the pelvis could be constructed, which would, to my mind, be far more accurate than any which have been produced by either of the two methods previously discussed. Such a schematic drawing as that to which I shall call attention later, certainly seems to be based upon a ground more worthy of credence, as a faithful representation of the pelvic relations, than a mere photograph of a section, into which probable grounds of error have unavoidably entered. It is not unworthy of the consideration of those who possess the means of recording large numbers of

[1] No two frozen sections can be found which agree in all particulars. Some present extremely wide variations; thus demonstrating the justness of my criticism.

[2] See articles of Schröder, Foster, Schultze, Martin, and others, mentioned in the bibliography.

carefully made measurements in their hospital practice, to begin at once such a series of experiments as shall tend to settle many points in anatomy as yet in dispute, and to place the knowledge of the topography of these parts upon a more scientific and reliable basis.

RELATIONS OF BONY POINTS TO EACH OTHER.—In studying the pelvic viscera, most of the points of controversy between authors of note as to the topography of the pelvic contents may be shown in an antero-posterior median section; hence, this is by far the most important view of the female pelvis. It is especially important, therefore, in endeavoring to construct a schematic diagram of this section, that the bony structures of the spinal column, sacrum, and pubes, be first represented as closely aspossible in accordance with the researches of Litzmann, Duncan, Braune, Pirogoff, Le Gendre, and Fürst,[1] before any attempt be made to depict the relative position of the pelvic organs. There seems to be little if any ground for adverse criticism of the drawings of Litzmann, which portray the normal relations of the bony points of the female pelvis (as deduced from a large and careful examination of many selected specimens); and his plates may well be taken as a model for this part of any schematic drawing. Since, to my regret, the limits of this article will preclude the discussion of the bony points of special interest in the pelvis, I quote for the benefit of the reader the admirable table of Foster in which he contrasts the measurements of some of the authorities above mentioned with those of Litzmann, in support of the accuracy of his own schematic drawing, made to illustrate the normal topography of the pelvic viscera.

Such a table is of value, moreover, in estimating, in any given case, the amount of deviation of the pelvic measurements from the normal standard.

---

[1] The plates of Savage, and the cuts found in Thomas, and most of the later works upon gynecology are *grossly incorrect* (in respect to the outlines of the sacrum and pubes) in representations of the sagittal pelvic section. In Savage's work, the sacrum in its thickest portion is ⅔ of the antero-posterior diameter of the superior strait; the tip of the coccyx lies far below the plane of the lower edge of the pubes, when it should be above it; and the sacral canal is inclosed by bone for its entire length.

of the Female Pelvic Organs.

MEASUREMENTS OF THE FEMALE PELVIS IN CENTIMETRES.[1]	Foster.	Litzmann.	Fürst. Average.	No. of pelves measured.
Promontory to upper border of symphysis pubis	11.7	11.4	11.7	19
" " nearest point "	11.0	11.0	10.6	19
" " lower border "	13.3	13.0	12.9	19
" " middle of third sacral vertebra	6.9	7.3	6.8	13
" " sacro-coccygeal joint	10.8	10.8	10.3	9
Sacro-coccygeal joint to tip of coccyx	3.8	3.9	3.3	10
Promontory to tip of coccyx	11.6	11.6	11.7	19
Highest to lowest point of symphysis	3.8	3.7	5.0	19
Upper border of symphysis to upper border of third sacral vertebra	13.5	12.3	13.5	16
Lower border of symphysis to sacro-coccygeal joint	12.5	10.8	10.9	9
" " " tip of coccyx	9.2	7.4	9.1	19
Upper " " middle of third sacral vertebra	13.9	12.7	13.7	13
" " " spin. process of last lumbar vertebra	18.8	....	16.9	12
Promontory to spinous process of last lumbar vertebra	7.6	7.6	6.0	12
" vertically to line from upper border of symphysis to spine of last lumbar vertebra	3.2	....	1.5	12
Tip of coccyx to horizontal line touching lower border of symphysis	2.0	....	2.4	19
Promontory to horizontal line touching lower border of symphysis	12.7	....	13.4	19
Horizontal distance of angle of sacrum from line falling vertically from promontory	7.5	....	6.9	19
Horizontal distance of middle of third sacral vertebra from line falling vertically from promontory	5.9	....	5.7	13
Horizontal distance of sacro-coccygeal joint from line falling vertically from promontory	7.3	....	6.6	9
Horizontal distance of tip of coccyx from line falling vertically from promontory	4.5	....	6.8	19
Middle of third sacral vertebra to sacro-coccygeal joint	4.5	4.4	4.7	8
Upper border of symphysis to that portion of it nearest to promontory	1.1	0.7	1.4	19

SITUATION OF POINTS OF SPECIAL INTEREST BETWEEN THE COCCYX AND THE SYMPHYSIS PUBIS.—It is absolutely essential, in order to properly construct a diagram of the topography of the pelvic viscera (*from measurements made upon the living subject*), that the soft tissues which intervene between the tip of the coccyx and the symphysis pubis be correctly represented in their relations to the bony outlines. It would

---

[1] The measurements given in the first column are those of Foster's original drawing; the *proportions* only are preserved in his reduced cut, originally published in this JOURNAL.

be a matter of less importance if these parts were not the subject of great differences of opinion between authors of note, and one which must be determined with the greatest care before the internal organs can be properly placed in a schematic drawing, in their normal relations to each other. There is perhaps no point in which antero-posterior median sections of the pelvis (as depicted in drawings generally

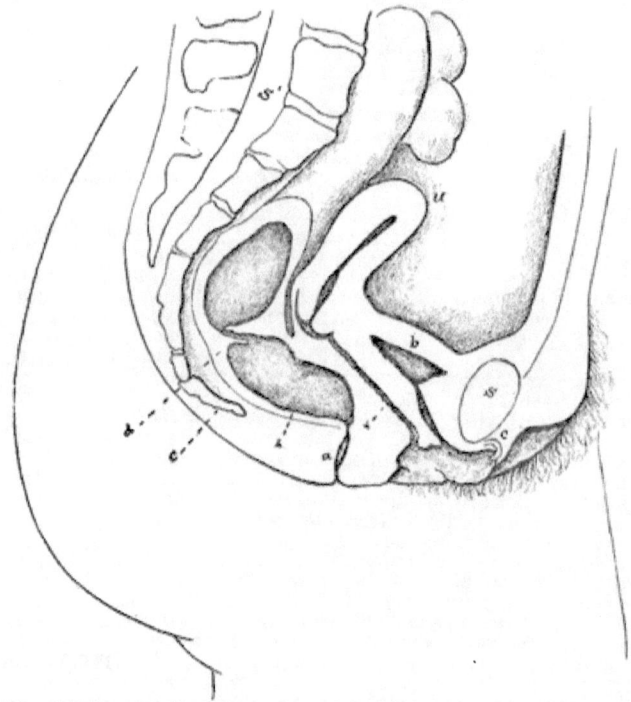

FIG. 1.—A diagram of the sagittal section of the pelvis of a living woman (modified from Foster), a, anal canal; r, rectum; v, vagina; c, clitoris; b, bladder, when collapsed; u, uterus; d, valves of the rectum (Houston); S, symphysis pubis; S', sacrum; C, coccyx.

accepted) show greater points of difference than in the region which embraces the anus, the perineal body, the various features of the vulva, the vulvo-vaginal orifice, and the meatus urinarius, as we pass from the tip of the coccyx forward. The relative situation of these parts, and the extent to which they project during life beyond the plane of the outlet of the pelvis, have been made a subject of special investigation by Schröder, Schultze, and Foster.

In the discussions which have taken place between the two authors first mentioned, as to the accuracy of their respective drawings, the frozen sections of Braune and Rüdringer were utilized by Schröder as conclusive evidence of the fidelity to nature of his own drawing. He claimed further that the accuracy of the drawing was sustained by the measurements which he had made upon the living subject.[1] It remained, however, for Foster, in an admirable paper published a few years ago, to demonstrate that the measurements of Schröder were faulty, from a defect in the method pursued by him; and to call attention once more to what he considered an improvement upon all previous methods of determining the normal relations of these parts to each other. The position of Foster was strengthened, some time before the appearance of the article referred to, by a similar criticism on the part of Schultze, who pronounced the method employed by Schröder as deective (so far as the deductions drawn concerning the projection of the soft parts beyond the plane of the pelvic outlet were concerned), from an oversight in the non-elimination of the thickness of the skin which covers the tip of the coccyx and the symphysis pubis.

It would add greatly to the value of this article if the methods employed by each of these three observers could be given in detail; but lack of space precludes more than a summary of their conclusions. Schröder places the extreme distance of projection of the soft parts beyond the plane of the pelvic outlet, in the nulliparæ, at 4.1 cm.; while Foster, on the other hand, considers 2.5 cm. as the average of this projec-

[1] Foster's measurements refer to subjects placed in the semi-prone posture, and Schröder's are not stated to have been otherwise made; this posture would tend to slightly diminish the extent of projection of the pelvic floor beyond the plane of the pelvic outlet from that which would exist in the erect attitude.

In pregnancy, *the pelvic floor projection is increased* in proportion to the weight of the enlarging uterus.

The "sagging" of the pelvic floor, observed in many of the frozen sections of the pelvis through its sagittal diameter (see those of Pirogoff, Fürst, Braune, and others), tends to support my view that death inevitably destroys to a greater or less extent the topographical relations of all the pelvic structures. Schröder has endeavored to justify his averages by measurements upon the frozen subject; a step manifestly illogical, since it is opposed by weighty objections. Foster and Schultze have probably placed the average more nearly in accordance with accuracy.

tion. Two separate tables of Schröder have been combined by Foster, to show the other measurements of this author, for the purpose of contrasting deductions made by himself, as the result of measurements on a much larger number of subjects by his improved method. I quote this compiled table of Schröder's measurements, in order to show the points of dispute between these two observers.

Measurements of Prof. Karl Schröder of Erlangen.	Distances from tip of coccyx to					
	Anus.	Fourchette.	Meatus urinarius.	Clitoris.	Lower border of symphysis pubis.	
	By tape measure.					By calipers.
	CM.	CM.	CM.	CM.	CM.	CM.
Average of the pregnant women	5.65	8.75	12.9	15.45	13.35	9.15
" " gynecological patients	5.4	8.3	12.0	14.5	12.6	8.29
" " nulliparæ	6.	9.	12.2	14.6	13.2	8.75

It will be perceived that the measurements made with the calipers (in the table of this author, as well as in those of Foster and Schultze) differ markedly from those made with the tape; since the actual distance between points in a straight line is determined by the former, and a section of the perineal curve by the latter method.

The *situation of the anus* from the tip of the coccyx, as the average of his measurements, is placed by Schultze at 5.9 cm.—33 women having been used as a basis of this deduction. Foster, however, places the average of 67 nulliparæ at 4.5 cm., and of 105 women who had borne children at 4.7 cm. from the tip of the coccyx. Both of these observers put their average distance below that of Schröder; yet the marked discrepancy between the average of Foster and that of the two other observers is due to the employment of the æsthesiometer in place of the tape; because the arc was measured by two, and the direct line between the points by one only. The method employed by Foster seems to me to be not only the most accurate, but also the best adapted for record upon a chart, or for use in the construction of a schematic drawing of the pelvis; hence

I quote the deductions, drawn by him from the same number of observations, respecting those other points between the coccyx and the symphysis where error is most likely to occur when making a chart, as regards their proper position and outline.

"The distance from the *anus* to the *fourchette* averages 2.7 cm. in nulliparæ, and 2.5 cm. in women who have borne children. Now, we have already seen that the distance from the tip of the coccyx to the anus averages 4.5 cm. in nulliparæ, and 4.7 cm. in women who have borne children. Expressed more graphically:

	Tip of coccyx to anus.	Anus to fourchette.	Tip of coccyx to fourchette.
"Average in nulliparæ,	4.5 cm. +	2.7 cm. =	7.2.
" " women who have borne children,	4.7 " +	2.5 " =	7.2.

"As to the other features of the vulva, I regard their precise situation as comparatively unimportant, and have therefore made but few observations bearing upon them. My impression is, that the upper or anterior extremity of the *vulvar fissure* is generally situated about opposite the middle of the symphysis pubis, or somewhat above this point.

"The *meatus urinarius* I have found in very few measurements to range from 0.7 to 2.2 cm. distant from the lower border of the symphysis pubis, and from 2 to 2.5 cm. from the fourchette in 'nulliparæ, and from 2 to 3.1 cm. in women who have borne children."

The same observer also states that his measurements showed that the *tissues about the anus* were the seat of the greatest projection from the line of the pelvic outlet. All of these measurements were made with the subject in the Sims' posture.'

[1] The researches of the three authors quoted above are a sufficient ground for adverse criticism of many drawings found in gynecological literature. They may also be used to sustain the author's position in reference to the non-reliability of frozen sections of the pelvis. A careful comparison of many such sections reveals the fact that the outline of the tissues which form the pelvic floor has been unavoidably distorted, either by cold or by loss of their muscular tonicity. Unusual stress is laid upon this point, because it is popular, of late, to decry any drawing which does not agree with some special frozen section that happens to coincide with the views of the critic.

By means of the data thus given, it is possible to construct a chart which shall represent the normal relations of the sacrum, coccyx, and pubis, in an antero-posterior median section of the pelvis, with a greater approach to accuracy, to my mind, than most of the plates now in existence exhibit; and to fill in the outline of the soft parts, which lie between the tip of the coc-

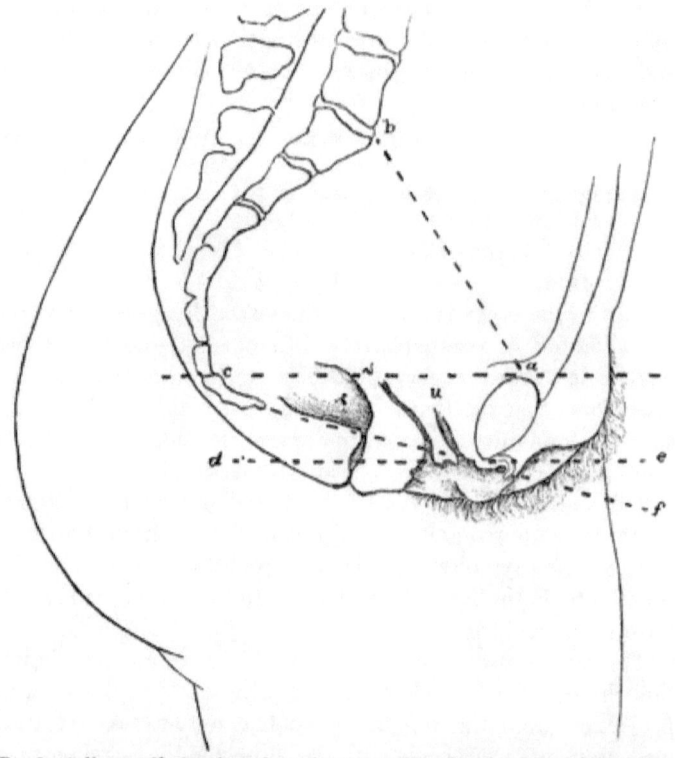

FIG. 2.—A diagram designed to show the planes of the female pelvis and the outline of the soft tissues which help to complete the pelvic floor. *a–b*, plane of pelvic brim; *a–c*, horizontal plane intersecting upper border of the symphysis; *d–e*, horizontal plane intersecting lower border of symphysis; *c–f*, plane of pelvic outlet; *u*, urethra; *v*, vagina; *r*, rectum.

cyx and the symphysis pubis, as they normally exist in the average Caucasian woman. A complete outline, not only of the bony walls but also of the soft structures at the pelvic outlet, may prove of great service in many ways. It will enable each observer to place the pelvic organs, in any case under his special notice, in their true relations to these important parts, provided proper measurements are carefully made and recorded; and it will also be of the greatest value as a basis for each observer in recording what his subsequent investigations lead him to believe to be the normal outlines of the organs which prominently appear in the antero-posterior median section of the female pelvis.

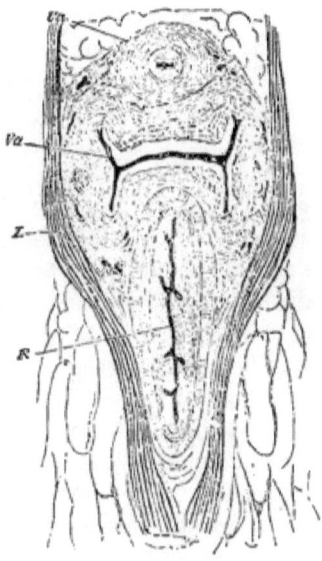

FIG. 3.—A horizontal section of the pelvic floor at the pelvic outlet above the level of the sphincters (Henle). *Ur*, urethra; *Va*, vagina; *R*, rectum; *L*, levator ani musle. The anterior and posterior walls of the vagina come together throughout.

THE PELVIC CANALS.—The transverse section of the pelvis (first devised by Henle to show the normal outline and relations of the three pelvic canals, viz., the urethra, vagina, and rectum) is now accepted as accurate. It shows that the vagina and rectum are not open tubes, but that they exist, when not distended, as mere slits in the pelvic section. The long axis of the vaginal slit is transverse, and corresponds, as Hart puts it, to the mouth of the woman in its general direction; that of the rectum is directed antero-posteriorly, thus forming a right angle with the long axis of the vaginal slit; while the urethra appears as a puckered and closed tube. These facts should be remembered in endeavoring to construct a schematic drawing of the antero-posterior median section of the pelvis, since the vagina must appear only as a line, while the rectum must occupy a broader space, because the long axis of the lumen, when it is closed, lies in the antero-posterior pelvic plane. The open anus, so commonly depicted in most of the gynecological contributions, is inaccurate, if the condition which exists during life is to be properly represented. This condition may exist in frozen sections of the pelvis,[1] but never during life, as any careful observer can attest, provided the sphincters of the rectum and the levator ani muscle are performing their proper functions.

THE PERINEAL BODY.—This structure, which has lately come into special prominence, forms a part of the pelvic floor, and must be drawn with its base looking downward in the erect posture of the woman, and lying between the anal orifice and the so-called "posterior commissure of the vulva." On antero-posterior median section of the pelvis, it presents, according to some authors, a triangular form, the apex of the triangle being directed upward, and interposed between the posterior wall of the vagina and the anterior wall of the rectum. It has been disputed by some that the tissue between the lower end of the vagina and the rectum has sufficient density to be properly considered as forming an apex to the perineal body, the point being made that, when two fingers are employed, one being inserted in the rectum and one in the vagina, no firm resistant body can be felt to exist between the fingers after they have each passed the thickness of the tissues which form the pelvic floor. This certainly agrees with my own observations made repeatedly upon the living female in the standing posture; moreover, many of the frozen sections of the pelvis fail to sustain the triangular shape of this body, as it appears to be rather quadrangular in form than presenting a distinct apex. Garrigues has compared its shape to the "cucurbit of an alembic."

[1] This is a further proof of the justness of my criticism of this method of studying pelvic topography. The tissues of a corpse can never be an infallible guide in deciding disputed points respecting the relations of organs during life.

The perineal body has been brought into prominence of late by an article which treated of its physiological functions, the accomplished author of which is certainly entitled to the consideration which has been paid by most reviewers to his researches in other departments. In this article, the perineal body is compared to the "key-stone of an arch," and its function stated to be a mechanical support to the remainder of the arch, viz., the posterior vaginal wall and the anterior wall of the rectum.

As the spirit of inquiry should always prompt an effort on the part of each observer to eliminate what seems to him an error, I may be pardoned if I endeavor to point out what seem to me to be the fallacies in the argument made in support of this theory of the wedge-like action of the perineal body.

In the first place, I would object (as the author of the paper seems to have anticipated that some one would) to the analogy drawn between this body and the key-stone of an arch. It is a principle of all mechanical devices into which a key-stone enters that its base must be directed toward, and not away from, the weight to be supported by it; otherwise, the keystone would slip out of its own weight and the arch would fall provided that it were not supported below. Again, no arch can have any sustaining power (in excess of the cohesiveness of the materials of which it is composed) provided that its concavity looks towards the pressure which it is designed to bear, unless it be supported below; in fact, no inverted arch is ever used under such circumstances without some firm support, as a straight beam would possess an inherent strength far in excess of an inverted arch unsupported. A careful study of nature, as revealed in anatomy alone, will convince any one that the simplest mechanical means are always used to accomplish any given end. If the shape of the perineal body be accepted as triangular (and it may well be considered as yet in doubt), and the muscular tissues which help to sustain it be considered as bearing an analogy to that of a supporting sling to the "keystone," the wedge-like action of this mass of elastic, fibrous, and muscular tissue may possibly be substantiated, provided that the later investigations of Hart, as to the physiological mechanism of the pubic and sacral segments of the pelvic floor, do not seem to the reader to utterly disprove it.

I am personally inclined to discard the wedge-like action of the perineal body as the true interpretation of its physiological function. This structure seems to be composed of a large excess of elastic tissue intermingled with the muscular fibres of the bulbo-cavernosus, sphincter ani externus, transversus perinæi, and ischio-coccygeus muscles, and also some fibrous tissue derived from the ischio-perineal ligaments, the deep layer of the superficial perineal fascia, the perineal septum, and the deep perineal fascia.[1] The result of the fusion of these structures is to produce a body which shall combine a great resistant power with a high degree of elasticity, two elements most essential to this part from its situation and the strain which it is called upon to bear in the act of delivery of the pregnant female. If, as I am inclined to believe, the " pubic segment" of the pelvic floor, as described by Hart, prolonged backward to the sacrum by the addition of the uterus and the sacro-uterine ligaments, is one great factor in supporting the pelvic organs and the weight of the super-incumbent structures of the abdomen, the perineal body will, of necessity, be deprived of much of its supposed physiological interest.[2] But this subject will be discussed in subsequent pages.

THE HYMEN.—In the vicinity of the vulvo-vaginal orifice may be seen the hymen. This thin septum of vaginal tissue[3]

---

[1] See article by the author, which treats of the female perineum, New York Med. Journ., July and August, 1882.

[2] To show that this view is shared by some of the later authorities, who have made a careful study of the anatomy of the female pelvic organs, I quote the following sentences from the late work by Hart and Barbour, which has created considerable attention: "Its functions (that of the perineal body) are important, but have been exaggerated and underrated. It gives a fixed point for many muscles, prevents pouching of the rectum forwards, and strengthens that part of the pelvic floor which has no posterior bony support." Moreover, the plates of Braune, Fürst, Hart, Pirogoff, Kohlrausch, Waldeyer, Henle and others which have been prepared from frozen sections of the pelvis, show extreme variations in its shape.

[3] The hymen, as shown by Budin, of Paris, is not composed of a reduplication of the mucous membrane of the vagina, but is a structure similar to that of the vagina itself. The vaginal orifice is therefore identical with the aperture surrounded by the hymen.

is not always present, but, when so, is attached, as a rule, to the posterior portion of the vulvo-vaginal ring. It most commonly assumes the form of a semilunar fold, with its concavity looking upward toward the meatus urinarius, and an opening is thus afforded between its free edge and the anterior vaginal wall for the escape of menstrual blood. There are a number of less common varieties which may be encountered. Among these may be enumerated the "annular form," where a central aperture exists in a continuous membrane; the "cribriform variety," where several smaller openings give it a sieve-like appearance; the "imperforate variety," which completely closes the vaginal orifice and occasions retention of accumu-

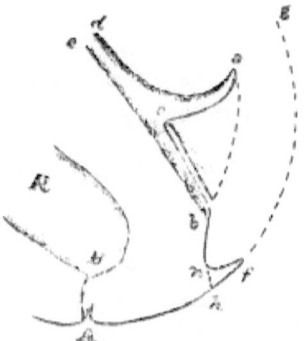

FIG. 4.—A diagram designed to show the normal position and attitude of the hymen (modified from Foster). *a–d*, anterior wall of vagina; *b–e*, posterior wall of vagina; *a–b*, attached border of hymen; *a–c*, free border of hymen; *b–c*, normal attitude of the hymen when the vaginal walls are approximated; *n*, fossa navicularis; *f*, outline of fourchette as seen when labia majora are pulled apart; *h*, dotted line showing position of the fourchette when the labia are in contact; *f–g*, outline of vulva; *A–h*, line of perineum; *A*, anus; *A–B*, anal canal; *R*, rectum. The vaginal walls are separated in this diagram to make the normal attitude of the hymen (when the walls of that tube are collapsed) more apparent. The space between the vagina and rectum is not present in nature, but is left in diagram to avoid confusion between the walls of the vagina and rectum.

lated menstrual blood; and the "fimbriated variety," which has been mistaken for a ruptured hymen on account of its fringe-like edge. This latter form has a medico-legal aspect, and its occasional existence should not be forgotten in examinations for suspected rape.

The hymen is not stretched tightly across the vaginal orifice, except when that canal is distended or its walls mechanically separated; on the contrary, it is usually sufficiently loose to lie closely in apposition with the posterior vaginal wall when that canal is closed. This fact can be verified by introducing the finger into the vagina (when the hymen may possibly not be detected—cases being on record where copulation and the delivery of a seven-month child has failed to rupture it) and then withdrawing it with the last phalanx flexed and in contact with the posterior vaginal wall, when it will usually be caught upon the finger tip and brought into view. It is not uncommon for an existing hymen to allow of the introduction of a speculum without laceration; and, doubtless, the presence of this membrane is often overlooked from a mistaken idea of its normal position. This septum has been compared to a "jib bellied with the wind," and crowded against the posterior wall of the vagina. Such a position of the hymen can be explained as the result of the collapsing of the vaginal walls when that canal is not distended, the posterior wall being brought into contact with the anterior by the muscular action of the pelvic floor.

It was first pointed out by Schröder that the "carunculæ myrtiformes," which are described by most authors as the result of rupture of the hymen, are not (as a rule, at least) the result of sexual intercourse, but rather of child-bearing. The pressure of the child's head as it passes over the seat of the hymen (possibly somewhat lacerated by coitus, but still present) creates a sloughing process in this delicate and poorly nourished septum, of which the isolated elevations upon the mucous membrane, called carunculæ, are the only remaining traces. Lusk states that in his examinations of young nulliparous prostitutes, present in hospitals, he has never found carunculæ myrtiformes; and I can add my own confirmation of this statement from a hospital experience confined largely to this class, and from subsequent observation.

THE VAGINA.—As has been previously stated, this canal appears as a transverse slit in that section of the pelvis devised by Henle to show the relative position and appearance of the three pelvic canals (Fig. 3). In the antero-posterior median section of the pelvis, it appears as a line only, and not as an open tube. It is often surprising, as well as amusing, to read the statement found in some of the standard text-books upon descriptive anatomy that the vagina follows and conforms more or less to the concavity of the sacrum. It is well known that the tip of the coccyx lies above the level of the symphysis pubis when the woman is in the erect posture,[1] and it is now practically admitted by most of the later authors upon this subject that the vaginal canal seldom reaches much above the level of the lower end of the sacrum. It cannot be said, therefore, to have much if any relation to the sacrum; since it does not extend to the limits of the bone posteriorly, nor lie, to any great extent, within its limits as far as the pelvic cavity is concerned. Its course, which is to my mind essentially a straight one, may be found to present curves, at times, from the effect

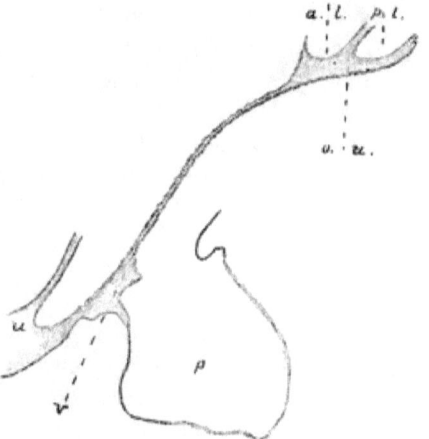

FIG. 5.—A diagram of the vaginal slit, as seen in a sagittal section, nearly life-size (slightly modified from Hart).
 u, urethra; v, vagina; a. l., anterior lip of cervix uteri; p. l., posterior lip of cervix uteri; o. u., os uteri externum; p., perineal body. The attitude of the cervix is distorted in the original frozen section, as in this cut.

of pressure, created by muscular action or by the distention of the bladder or the lower part of the rectum. Some authorities describe a normal curve in the lower part of the vaginal canal, and Thomas attributes to it a double curve, resembling the letter S, to which he believes that some of its abnormalities after a laceration of the perineal body are due; but I am inclined to think that all vaginal curvatures are the result to a greater or less extent of the action of the muscles which help to form what

[1] This fact would not be sustained, if many cuts incorporated in gynecological literature were to be taken as our guides in anatomical study.

Hart designates as the "sacral segment" of the pelvic floor, since the canal appears to be nearly straight in all sections of the corpse and also in subjects where the perineum is extremely relaxed as the result of marked debility. Such muscular action would tend to crowd the posterior wall of the vagina upward and forward, and thus help to produce a curve which would be wanting if the muscles were relaxed. There is every reason to believe that a certain degree of muscular tonicity in the perineum is present in every living female, if the pelvic floor is intact and the subject in good general health; in fact, the "feel" of the perineum, when the patient is in the erect posture, has been utilized by some of the later observers as a guide of value in determining the existing state of general health—a relaxed condition being an indication of a state of debility. This muscular tonicity conduces toward a close approximation of the pubic and sacral segments of the pelvic floor; hence it is easy to account on this ground for the curve given to the vaginal canal in the drawings of Thomas and Foster, and to explain the absence of any marked curve in the plates of Savage, Fürst, Braune, Pirogoff, and others which represent the condition found either by dissection or in frozen sections of the body.

A late monograph of Hart, to whom a gold medal and the Syme fellowship were awarded for the excellence and originality of his investigations, states that that line in the antero-posterior median section of the pelvis, which marks the situation of the vaginal canal, is a boundary line between what he designates as the "pubic" and "sacral" segments of the pelvic floor. Under the former term, he embraces the structures which lie anteriorly to this line, viz., the anterior vaginal wall and the contiguous wall of the bladder; while, under the latter term, he includes the strong muscular partition which incloses the pelvic outlet, the perineal body, and the posterior wall of the vagina. These two segments, according to this author's description, overlap each other like valves. The pubic segment, being attached anteriorly and unattached behind, can be lifted over the object to be expelled through the vaginal canal; while the sacral segment, being attached behind and movable anteriorly, can be pushed backward in proportion as the vaginal canal is to be distended. Between these two segments (the "pubic" rising, and the "sacral" being depressed or crowded backward) the fetus is expelled. In the normal state, these two segments so overlap each other as to present a pelvic floor which is practically unbroken. The power of sustaining the uterus is attributed by this author entirely to the sacral segment. He furthermore states that the vaginal canal makes an angle of 60 degrees with the horizon; and that its walls are in close apposition with each other, in all possible positions of the body. They may become separated from each other, however, if artificially opened during the genu-pectoral position, because air is then sucked into the vaginal canal; and also during the passage of a fetus, menstrual blood, etc., from the cavity of the uterus, or the introduction of the finger (during a digital examination) or of a speculum.

It has been stated, in an admirable review of this work,[1] that the author has lost sight of one important fact in connection with the pubic segment, viz., that the sacro-uterine ligaments are a practical extension of this segment to the posterior bony wall of the pelvis, thus transforming it into a supporting medium for the uterus, irrespective of the sacral segment. This seems to me to be a point well taken; and I would suggest that the pubic segment be made to include not only the anterior vaginal wall, and the contiguous portion of the wall of the bladder, but also the uterus itself and the sacro-uterine ligaments, since the latter are abundantly supplied with muscular fibres and must of necessity possess a contractile power. Another important point seems to me to have been lost sight of in this description, namely, that the sacral segment is continued as far forward as the symphysis pubis in all antero-posterior sections of the pelvis *except in the median line;* here the opening of the external genitals exists, and it apparently makes its termination at the so-called posterior commissure of the vulva. This anatomical fact is not made apparent in Hart's drawings of this segment, and must, to my mind, destroy all similarity of this segment to a hinged-flap, as the author would lead his readers to accept.

Instead of two flaps, the more accurate conception of the pelvic floor, as it seems to me, would be to regard it as com-

[1] N. Y. Med. Jour., Sept., 1881.

posed of two partitions nearly parallel with each other, the dividing line between which is the vaginal canal; the upper partition comprising the pubic segment of Hart prolonged backward to the sacrum by the muscular tissue of the uterus[1] and the sacro-uterine ligaments (which are rich in muscular tissue), while the lower partition would include the structures of the perineum proper, held in close apposition to the upper partition by the natural tonicity of the levator ani and transversus perinæi muscles, and the elastic and muscular fibres found in the tissue of the sub-peritoneal pelvic space.

With these slight modifications, the essay of Hart seems to have presented the physiological construction of the female pelvic floor more in accordance with my own conception of the parts than that of any other author. The fact that the cervix can be drawn down to the vulva, without any great amount of force being used, does not seem to me to exclude the sacro-uterine ligaments from being considered a part of the upper segment; nor does it prove that they are not important agents in supporting the uterus, since their anatomical construction is such as to admirably adapt them to yield to a sudden strain and easily regain their proper tone. That we really have some such supporting partition, in addition to the muscular structures of the perineum and the perineal body, can easily be demonstrated on the living subject by an experiment which has been suggested by another, but which I have personally verified over and over again. If the perineum be pulled downward with a Sims' speculum, when the woman is in the standing posture,

---

[1] This view was first, to my knowledge, advanced in print by Foster (Gynecol. Trans., 1881), who expresses himself as follows:

"Except to resist extreme displacements of the organ, the broad ligaments, the round ligaments, the bladder, the rectum, and the perineum take no part among the supports of the uterus. Ordinarily, the latter consists wholly of the anterior wall of the vagina in front and the sacro-uterine ligaments behind, which together constitute what may be termed a beam traversing the pelvis antero-posteriorly, on which the uterus rests, being interposed between them, firmly attached to the one anteriorly and the other posteriorly, making of them, so far as the mechanical effect is concerned, one structure. Thus, and to this extent only, is the vagina —and then only its anterior wall—one of the much-debated supports of the uterus. It is not a prop; still less does it support the uterus by virtue of any rigidity of its own *as a tube*, as has been claimed (the mechanism being likened to that by which a bough sustains its fruit), but its anterior wall acts simply by reason of its connections with other parts."

the uterus does not fall to any appreciable extent, although the weight of the super-incumbent organs of the abdomen is exerting its greatest influence toward a downward displacement; this would not be the case if the structures which support the uterus were confined entirely to the perineum itself, and it seems conclusive proof at least of the valuable aid afforded by some other structures. If these modifications of Hart's views be accepted, the act of expulsion of the fetus, or of some abnormal growth, from within the cavity of the uterus

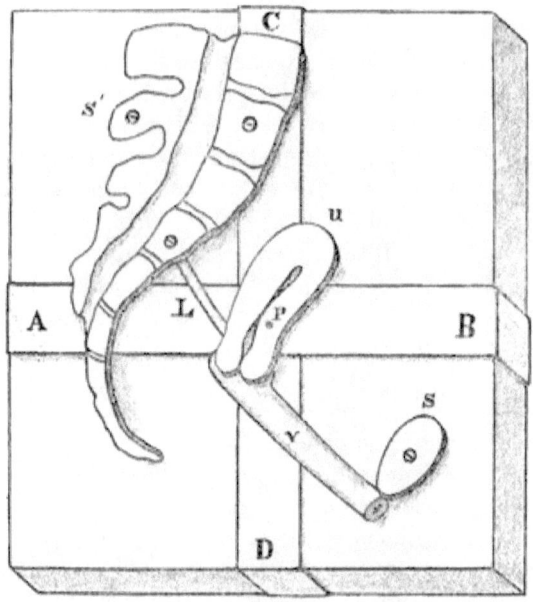

FIG. 6.—A diagram of a model designed to show the upper supports of the uterus and their relations to the mechanical treatment of uterine displacements.

A B, and C D, two elastic bands which intersect at the point of attachment of the uterus (U), which is fastened to them by a pivot (p); S, symphysis pubis, fastened firmly to block; as is also the sacrum (S'); U, uterus, made movable in all directions by elastic bands (A—B and C--D); V, a rubber tube, to represent the vagina; L, a rubber band, to represent the sacro-uterine ligaments.

is associated with the piercing of two septa; rather than the folding back of one flap and the raising of another, both of which are presumed by Hart to be unattached at one extremity. There has been a working model devised by my friend Dr. F. P. Foster, which illustrates in a most complete and simple way the upper support to the uterus and some other points of interest, among which may be mentioned the action of the Hodge pessary upon retroversion of the uterus. The probable mechanism of this instrument, as far as my researches go, was first explained by Peaslee and Bantock. As the subject has been discussed, however, by many authors, and various explanations given of the true mechanism of this pessary, I deem it of sufficient importance to warrant the description of this model and the insertion of a few scattered hints which are suggested by it.

This drawing represents the bony outline of the sacrum and pubes sawn out of wood and placed immovably upon a board in their proper relations with each other. A piece of wood is also similarly sawn into the proper shape of the uterus and attached to the board by two broad strips of elastic, which cross each other at the point of junction of the uterine cervix and body; these strips being so fastened to the model of the uterus as to allow of its motion in every direction, while they tend also to restore it to its proper place when traction ceases. The elastic bands of the model may be considered as analogous to the elastic and muscular fibres which exist (1) in the cellular tissue of the subperitoneal pelvic space and (2) in the ligaments of the uterus itself. The cervix of the modelled uterus is now inserted into the open mouth of a rubber tube (analogous to the vagina) and the anterior wall of this tube is drawn tense and fastened to the lower border of the symphysis pubis. The anterior end of this rubber tube is cut off at the symphysis to correspond to the mouth of the vagina. Finally, the uterus is connected to the sacrum by an elastic strap which is made continuous with the point of attachment of the rubber tube to the posterior lip of the cervix; this represents the sacro-uterine ligaments and completes the upper segment of the pelvic floor, which is thus made to consist of the anterior vaginal wall, the uterus itself, and the sacro-uterine ligaments. As the wall of the bladder is in part continuous with the anterior wall of the vagina, the model is to all intents and purposes a perfect representation of the conditions which exist in nature. Now, if the posterior vaginal wall be crowded backward in this model by the introduction of anything into the vaginal canal (which lies tightly closed in the model on account of the tension exerted upon its anterior portion) the body of the uterus will be seen to tilt forward in exact proportion to the amount of pressure created upon the posterior wall of the vagina. This is believed to be the true mechanism of the Hodge pessary, and the opinion of Bantock seems to be confirmed by the researches of Hart who publishes a drawing, as a diagrammatic representation of the action of this support. Some other hints are afforded us by this ingenious model, which cannot be properly discussed here, but which may be simply alluded to, viz., that the uterus can be drawn downward and again resume its normal position, by the aid afforded by the anterior wall of the vagina and the sacro-uterine ligaments; that the elastic and muscular fibres of the pelvic cellular tissue and the uterine ligaments are important agents in enabling the pressure exerted upon the posterior vaginal wall to create anterior deflection of the body of the womb (since they practically form an elastic support to that organ binding it to the lateral wall of the pelvic cavity and thus enable the organ to respond to pressure, as a lever with a fulcrum in its middle point would do); and, finally, that the anterior vaginal wall becomes so closely intermingled with the muscular fibres of the uterus and the sacro-uterine ligaments as to constitute what may be practically regarded as one continuous structure.

The *length* of the vagina is usually greatly overestimated, on account of its excessive distensibility. While marked variations may be noticed in different subjects during life, due chiefly to alterations in the position and weight of the uterus, the average length of the anterior wall may be given as about two and one half inches and that of the posterior wall as a trifle over three inches. The posterior wall is longer on account of its attachment to the posterior portion of the cervix, upon which it is prolonged to a greater extent before its attachment, than in the case of the anterior wall.

The anterior wall of the vagina is loosely attached to the wall of the bladder by connective tissue in its upper half. In its lower half, however, it is intimately associated with the tissues about the urethra; the partition which is thus formed being called the "*septum urethro-vaginale.*" The posterior wall of the vagina is separated from the rectum, in its upper fifth, by the peritoneal pouch, the "*cul-de-sac of Douglas;*" while "the perineal body" is interposed between it and the rectum, for an extent whose limits are not as yet clearly defined, in its lower portion. In the space intervening between the perineal body and the pouch of Douglas, the vagina and rectum seem to possess a degree of mobility (to allow of their ever-varying degrees of distention) which rather tends to refute the statement made by some authors as to an intimate blending of these tissues into the so-called "*septum recto-vaginale.*" Braune states that these organs are separated by *two layers* of fascia and loose connective tissue, in this locality. In support of this view, a case related to me by A. H. Smith, M.D., of New York City, possesses special interest. A woman was admitted to one of the hospitals and placed under his care. She presented a hematoma as large as an orange between the rectum and vagina in the *lower third* of the space, immediately above the level of the pelvic floor. This unique case clearly demonstrates that the vagina and rectum do not become so closely united as to form the so-called "vagino-rectal septum," as blood could not have become effused to any appreciable extent in the locality mentioned without the walls of the vagina and rectum were separated by connective tissue or fasciæ. A case somewhat similar to that reported above is recorded in Emmet's recent work on gynecology, in which a pelvic hematoma separated the vagina and rectum as far downward as the level of the pelvic floor. The existence of fasciæ between these organs as far down as the level of the pelvic floor I have been able to personally verify at the dissecting table.[1] After a complete removal of the structures which enter into the formation of the female perineum, the vagina and rectum may be torn apart with the greatest ease and with no apparent injury to the walls of either tube.

As has been stated in previous pages, my own researches tend to discredit the penetration of the perineal body, between the rectum and vagina, above the limits of the pelvic floor.

The *direction* of the vaginal canal, in the upright posture, has been variously represented. Most of the drawings of the later investigators place the axis of the vagina as more nearly vertical than my own observations would tend to confirm. Even Hart's computation, that the axis of the vaginal canal forms an

[1] In one frozen section, the pouch of Douglas descends almost to the ostium vaginæ.

angle of sixty degrees with the plane of the horizon (presumably in the erect posture) seems to me to be excessive. I am inclined to believe that forty degrees with the horizontal plane would more nearly indicate the normal angle made by the vaginal axis in the erect posture,[1] and sixty degrees when the woman is placed upon the back. The introduction of the finger is not a safe guide to determine this point, as pressure may be made unwittingly upon either the anterior or posterior vaginal wall. Several ingenious instruments have been devised, which may be used in determining this point. There is scarcely any doubt, however, that the vaginal canal is more nearly horizontal, when the woman is in standing posture, than is generally supposed; and, if some existing cuts in anatomical and gynecological works are destined to be perpetuated, the error is liable to be eradicated slowly.

Points of special Interest pertaining to the Vagina.—Both the anterior and posterior walls of the vagina are triangular in shape, the base of the triangle being above. They are united at their sides.

Note that in sagittal pelvic sections, the anterior wall appears almost straight, while the posterior wall presents curves which vary with the distention of the rectum, the position of the uterus, the amount of traction upon the vagina by the sacro-uterine ligaments, and the tonicity of the pelvic floor. Distention of the bladder may also create an alteration of the vaginal axis, on account of a backward displacement of the uterus which is thus created.

Most of the cuts in anatomical and gynecological works represent the vagina as an open tube—a gross error which is now universally acknowledged, but which is repeated, in order, as it were, to let the student *see* the vagina. Hart humorously remarks, in discussing this point, "It is no more necessary to figure the vaginal walls always apart, than it would be to always sketch a man with his mouth open in order to render it visible."

If the finger be passed to the lateral aspect of the fornix of the vagina in the nulliparous woman, it will lie in contact with the *base of the broad ligament* of the uterus—a point which may often be useful in diagnosis.

In the genu-pectoral posture, the vaginal walls tend to separate as soon as air is admitted by the introduction of the finger or a speculum. This fact, which is also true of the semi-prone posture, explains the utility of the Sims' speculum, as well as that of other modifications of the same. In addition to this alteration from the normal state of the vagina, there will be noticed, as a result of this posture, a marked increase in the length of the walls of the tube, an approximation of the anterior wall to the posterior aspect of the symphysis pubis, and a similar displacement of the posterior wall of the vagina toward the cavity of the sacrum.

The sphincter vaginæ muscle consists chiefly of pubo-coccygeal fibres of the levator ani muscle. There is no anatomical basis for the belief that the bulbo-cavernosi muscles exert an influence upon the vagina; hence the name "sphincter vaginæ" is improperly applied to them (see article by author on the Female Perineum, *N. Y. Med. Jour.*, August, 1882).

Pressure made upon the posterior vaginal wall in the region of the fornix vaginæ, so as to carry it toward the sacrum, tends to increase the anteversion of the uterus, provided that organ be not retroflexed. If the fundus be fixed, the state of retroflexion of the womb is increased by pressure made at the fornix vaginæ in the antero-posterior axis. If pressure be made in the anterior portion of the fornix vaginæ, the tendency of the force so applied is to draw the cervix forward and to rotate the normal uterus backward, unless the organ happen to be flexed; in the latter condition, the flexion is not apparently modified. The effects of traction thus made upon the vaginal walls are most apparent when the patient is in the Sims' posture. These experiments are of value as tending to prove that no vaginal pessary is capable of relieving anterior or posterior flexions of the uterus, because the body of the organ is only acted upon in an indirect way through its union with the cervix.

The relations of posture of the patient to examination and treatment is thus summarized by Hart and Barbour:

"The *side-lateral*, where the patient lies upon her side in the ordinary way, is convenient for vaginal examination; passage of Fergusson's, Neugebauer's, or Cusco's speculum; passage of the sound or catheter.

"The *dorsal posture* is imperative for abdominal examination and the bi-manual.

"The *semi-prone* is the best posture for the passage of Sims' speculum; vesico-vaginal fistula operation.

"The *lithotomy posture* is especially valuable for operations on the perineum, vaginal walls, cervix, and uterus.

"The *genu-pectoral posture* is useful for replacement of the retroverted uterus."

Pawlick, of Vienna, has pointed out that the *genu-pectoral posture* is useful in catheterization of the ureters. His method will be described later.

The vagina derives its blood-supply from the hypogastric, uterine, vesical, and pudendal arteries. During gestation, the pulsations of the uterine artery may usually be felt through the vaginal walls at its upper portion. This is considered by most authorities as a valuable inferential sign of that condition.

The vagina is inclosed by a plexus of veins that are destitute of valves, and which anatomose freely with the veins of the perineum, and also with the plexuses which surround the neighboring pelvic viscera. The absence of valves renders venous stasis apparent in the vagina, as evidenced by a deep purple color of the mucous lining, when pressure from any source impedes the return circulation. Jacquemin and Kluge have urged that this peculiar color of the vagina (compared by them to wine-lees) be included among the early symptoms of pregnancy, because it is commonly produced by the pressure created by the descent of the gravid uterus in the early months. It may, however, be present in prolapse of the uterus, pelvic tumors, etc. The free communication of the vaginal veins with those of neighboring organs, and the contiguous perineal structures, renders a disturbance in any one of these parts liable to a simultaneous disturbance in the circulation of the others.

The vaginal walls occasionally become prolapsed. This is more common during pregnancy, because a marked hypertrophy of the vaginal walls exists at that time. It may also be due to causes which impair the normal tonicity of the anterior vaginal wall or the sacro-uterine ligaments. As a consequence of prolapse of the vagina, a displacement of the uterus and bladder from their normal relations is almost unavoidable. Some authors claim that the mucous coat may be prolapsed, independently of the other coats of the vagina.

The great distensibility of the vagina should not be overlooked in plugging it, as the functions of the bladder and rectum may be seriously embarrassed by pressure.

Cystic tumors, apparently connected with the vagina, are often produced by distention of urethral follicles.

Preternatural dilatation of the vaginal orifice leads, sooner or later, to vaginal invagination or prolapse of the uterus.

Destruction of the "perineal body" leads to vaginal and rectal prolapse and subsequent uterine displacements.

Opening of pelvic abscesses through the vaginal walls is sometimes associated with fatal hemorrhage.

The whole thickness of the vaginal tissues may be accidentally removed during surgical procedures for the relief of cysto-vaginocele and recto-vaginocele; because the points where such protrusions are most common are situated where the vaginal walls are scarcely more than two lines in thickness, and they may possibly be much attenuated. Huguier has demonstrated, however, that this mistake has little if any danger, and forms rather an integral part of the success of the operation devised by him.

THE RECTUM.—This organ, situated between the vagina and the anterior surfaces of the sacrum and coccyx, should next engage our attention in studying the topography of the pelvic organs. It does not lie in the median line of the pelvis for its whole extent, being deflected, in its upper portion, toward the left side.[1] It is not sufficient, therefore, for the thorough study of its topographical relations that an antero-posterior median section of the pelvic structures be alone examined. Yet, as this particular section is the most important of any of the pelvis because it affords the most satisfactory view of the topographical relations of the various organs to each other, it seems particularly desirable that it should be correctly represented—a desideratum not often found. Most of the cuts incorporated in the works of the greater anatomists, and, I regret to say, also in those dealing exclusively with the gynecological department, represent the rectum as piercing the muscular struc-

---

[1] The rectum extends from the left sacro-iliac synchondrosis to the anus. Its first portion curves downward, backward, and inward to the level of the third sacral vertebra and is completely invested with peritoneum. About three inches from the vaginal orifice, the peritoneum leaves the walls of the rectum to form the anterior layer of Douglas' pouch.

tures of the female pelvic floor obliquely and reaching to the skin; while the anus is usually depicted as open, affording no apparent obstacle during life to the escape of its contents. It certainly needs no lengthy argument to convince any thinking mind that this cannot be, and is not, the normal state of the rectum during life. If the finger be passed into that canal of the female in the erect posture, it will be found that the sphincters create no slight resistance to the act. It will also be perceived, if the observations of others agree with my own, that the thickness of the pelvic floor in this region is about one inch (rather in excess of than under it); and that the anal canal perforates the muscular structures which form the pelvic floor in a direction nearly vertical, but slightly forward of the vertical line rather than behind it.[1] This canal will be found to open into a closed cavity (the lower part of the rectum), because the walls of the tube are compressed from the sides as shown in the cut of Henle (Fig. 3). The direction of the axis of this cavity will compel the finger to be now passed almost directly backward toward the sacrum. The lower part of the rectum is found to be empty, as a rule, because the feces are commonly retained higher up, until the act of defecation is demanded.[2] The great breadth of the rectum, which appears in all antero-posterior median sections of the pelvis, is to be explained rather by its method of collapse (the long axis of the lumen being parallel with the plane of the section) than by any misconceived idea that it is a distended tube, as the shading in most of the cuts published would seem to justify. From the internal opening of the anal canal to a point which corresponds to the level of the os uteri externum, the rectum runs backward and slightly upward; here a constriction may be detected which is described by Luschka as "the third sphincter of the rectum," although Nélaton was probably the first to apply the name and describe the existence of such a muscle. In an excellent article by Kelsey upon this anatomical feature of the rectal canal, the literature is exhaustively reviewed, and the probable nature of this constriction as well as its physiological

[1] Hart and Barbour put it as forming a right angle with the vaginal axis.
[2] Extreme distention of the rectum may cause the vaginal axis to be displaced forward so far as to be almost vertical.

functions are ably stated. Whether this so-called "third sphincter" consists of a muscle or of folds of mucous membrane, and whether it possesses any definite situation and attachments in either case, seems to be a matter of controversy between such eminent authors as Sappey, Hyrtl, Velpeau, Nélaton, Pétrequin, Houston, O'Beirne, Chadwick, and many others who have investigated the subject. To discuss the pros and cons of this matter, or to enter into the physiology of the act of defecation with sufficient minuteness to explain the views advanced as to the probable mechanism of this constriction, would exceed the limits of this article. The reader is referred to the bibliography appended to it for more definite information, and especially to the article of Kelsey, where all the views are contrasted.

Hart, of Edinburgh, has lately advanced some ideas concerning the method of collapse of the rectal walls which seem to me to be fallacious. He states that there is absolute proof that the rectum collapses, like the vagina, by an approximation of the anterior and posterior walls, as the result of intra-abdominal pressure. The proof which he adduces is a clinical one, viz., that vaginal pessaries would otherwise be forced out during the efforts of defecation. Now this is just what does occasionally occur when the perineum has suffered laceration, or is in a state of subinvolution; and Henle's cut (Fig. 3) points clearly to an approximation of the lateral walls. I do not believe that intra-abdominal pressure has anything to do with the collapse of the lumen of the rectum. There is elastic tissue enough in the sub-peritoneal portion of the pelvis to act as lines of traction upon both the vagina and rectum (Savage and others); and the upper wall of the vagina assisted by the sacro-uterine ligaments, is capable of resisting all ordinary pressure from above. The views of Hart himself would seem to sustain this latter objection. Again, the valves of Houston would seem to be a natural barrier to the effects of intra-abdominal pressure upon the contents of that portion of the rectum which lies below them.

The author quoted above believes that the valves of Houston assist in preventing eversion of the mucous membrane of the rectum, by the contraction of sphincter-like fibres during the act of defecation, while they assist at the same time in the expulsion of fæces.

It may be stated, I think, with an approach to accuracy, that a constriction does exist in the rectum at a point situated about four inches distant from the anus,[1] and that feces are seldom found in the portion of the rectum which lies below this constriction. The condition, which I believe exists most commonly, as tending to explain this constriction which may be appreciated during life, is the presence of two or three folds of mucous membrane which overlap each other like valves. These were believed by Houston, who first described them, to be so placed as to act as a mechanical arrangement to prevent the feces from forcing their way to the neighborhood of the anus where their presence would create a desire to defecate. The investigations of Rosswinkle, Hyrtl, Henle, Sappey, and Kohlrausch,[2] sustain the investigations of Houston, as regards the existence of these valves, although they differ among themselves as to the number present and their exact seat.

The fact that distention of the lower part of the rectum with feces may modify the position of the uterus, and also crowd the upper two-thirds of the vagina forwards, is well recognized by all gynecologists. I am inclined to argue, however, that marked displacements of the uterus from this cause are less common than are usually supposed; because the accumulation of fecal matter is normally above the point of constriction in the rectum, while the lower part of that tube is usually empty except during the act of defecation.

The close proximity of the rectum to the posterior vaginal wall and the uterus, coupled with the fact that this tube possesses a high degree of distensibility, has brought rectal exploration into prominence as one of the means of determining the existing conditions of the vagina, uterus, ovaries, the rectum itself, and even of the kidneys. Prof. Simon, of

---

[1] Other valve-like forms exist, at about one and a half inches from the anal canal; and, in the vicinity of the sacral promontory, oblique folds are very apparent, as a rule.

[2] This most excellent monograph, which is most rare, shows a superbly prepared section of a frozen corpse reproduced by an entirely novel process. The valves of the rectum are fortunately made very apparent. I am indebted to Dr. Bullard, of New York, for the use of his copy. Few, if any, reproductions of this plate give a proper conception of its beauty

Heidelberg, has done much to bring this step before the notice of the profession. The constriction spoken of as existing in the rectum does not seem to seriously interfere with the introduction of the fingers beyond it, after the hand has been introduced through the dilated anus. The uterus can be drawn down easily within the complete grasp of the fingers when necessary, after the hand has been thus introduced within the rectum, and small tumors of the fundus can often be detected by this procedure. Rectal exploration is not entirely devoid of danger to the patient; and a serious objection to its use is that the hand of the explorer (unless it be extremely small) is frequently so compressed by the rectal walls, that freedom of motion of the fingers is greatly interfered with. Frequent examinations of this character are liable to result in laceration of the walls of the rectum, and a permanent paralysis of the sphincter muscles of the anus may also be induced. The later works upon gynecology mention fatal results which have followed this procedure; hence it is one which should be resorted to only in cases of extreme necessity.

The *sacro-uterine ligaments* (folds of Douglas), which have been referred to as one of the possible factors in assisting the anterior vaginal wall to support the uterus, pass around the rectum upon either side, in order to bind the structures with which they are intimately connected in front—the muscular fibres of the anterior vaginal wall, and the uterus itself—to the second bone of the sacrum.[1] This point can be best studied from a superior view of the pelvis and its organs. The construction of these ligaments has been referred to in a previous page.

The rectum is incompletely invested with peritoneum at its upper part; the middle portion is only partially covered, the peritoneum gradually passing from the posterior surface and sides of the tube; the lower portion has the pouch of Douglas interposed between it and the vagina. This pouch will be described later.

Points of special interest pertaining to the rectum.—The close proximity of the rectum to the uterus, vagina, pouch of Douglas, broad ligaments, and ovaries renders it a valuable

---

[1] During gestation, the points of bony attachment of these ligaments rise gradually until they reach the level of the sacral promontory.

means of perfecting diagnosis. In addition to Simon's method of rectal exploration, less severe and dangerous steps are often employed. These may be enumerated as follows: (1) Digital touch; (2) conjoined recto-vaginal touch; (3) abdomino-rectal method; (4) abdomino-recto-vaginal method.

In all of these procedures, the rectum should be first emptied by an enema. In some cases, a purgative should precede the enema; being administered at night, previous to the proposed examination.

The simpler method (*digital touch*) is commonly employed

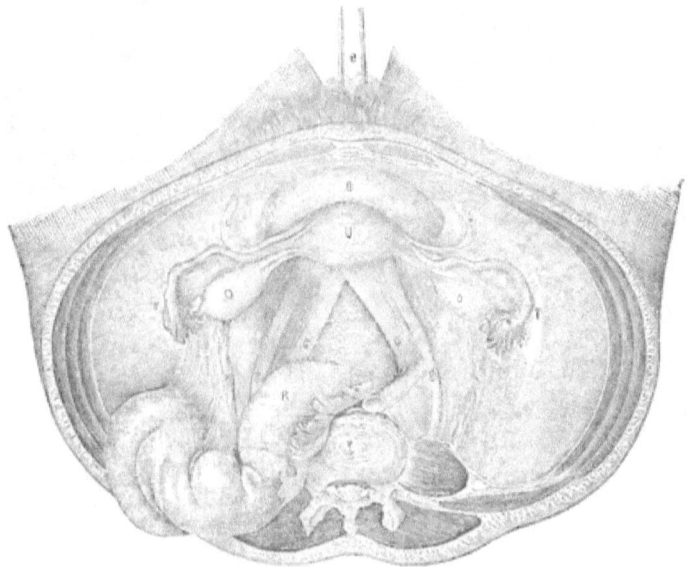

Fig. 7.—Superior view of the pelvis and its organs (Savage). B, bladder; U, uterus drawn down by vulsellum; T, Fallopian tubes; O, ovary; R, rectum; V, sacrum; C, sacro-uterine ligaments; L, round ligaments; *g*, ureter; *e*, loop attached to uterus; *a*, spermatic vessels, often prominent under their peritoneal covering.

in the virgin. It is useful in detecting hemorrhoids, fissures, rectal stricture, rectal polypi, ulcers, and malignant growths. The cervix can be felt through the anterior wall of the rectum. This may be mistaken for the body of the uterus on account of its apparent length—since it seems longer than the more common vaginal touch would lead an inexperienced examiner to suspect. The normal ovaries may be felt, as small oval-shaped bodies which are tender to pressure, by pushing the finger high up against the lateral walls of the rectum. A retroverted uterus may often be completely examined. In cases of marked acute flexion, the finger may sometimes detect the angle of flexion, and feel the fundus passing forward from it. Abdominal pressure greatly assists the finger in the rectum in detecting abnormalities of the uterus.

The *volsella* may be employed with advantage in connection with rectal exploration. This instrument has a decided advantage over abdominal pressure in cases where the walls of the abdomen are rigid. In this way, the mobility of the uterus may be determined; the utero-sacral ligaments can be felt as tense cords if involved in cicatricial tissue; the posterior surface of the uterus can be felt; the presence of fibroid tumors may be diagnosed; the ovaries are made more accessible; and, finally, the length of the uterus can be accurately determined.

The *recto-vaginal* method is performed by passing two adjoining fingers (the first and second are preferable) simultaneously into the vagina and rectum. In this way, the anterior wall of the rectum and the posterior wall of the vagina are included between the fingers. It may be used conjointly with abdominal pressure (the *abdomino-recto-vaginal* method). It is useful in the detection of retroversions, anteflexions, prolapsed ovaries, tumors of the rectal or vaginal walls, circumscribed effusions of blood etc.

Finally, injections of air into the rectum will distend it at high as the sigmoid flexure, and often prove a valuable adjunct to rectal exploration by any of the methods previously described.

The "perineal body" prevents pouching of the anterior wall of the rectum. It may also assist in deflecting the feces through the anal canal, whose axis forms a right angle with that of the rectum when the sphincters are contracted. Thomas attributes to this body the function of a support to both the anterior rectal and posterior vaginal walls (see previous pages of this article).

The levator ani muscle has the following physiological attributes: 1. It strengthens the deep sphincter of the rectum. 2. It helps to restore the inversion of the rectal mucous membrane after the act of defecation is accomplished. 3. It supports and raises the rectum and pelvic floor by means of fibres which are closely intermingled with the external sphincter muscle of the anus and the longitudinal fibres of the rectum. 4. It acts on the tip of the coccyx, which it tends to draw forward, and thus to increase its curvature. 5. It is enabled to sustain pressure in excess of its apparent strength by the toughness and elasticity of the recto-vesical and levator ani fasciæ, between which its fibres are placed. 6. It is the physiological antagonist to the diaphragm. 7. It acts as a sphincter muscle to the vaginal orifice by its longitudinal and sling-like fibres, and, possibly, compresses the urethra.

The longitudinal fibres of the rectum act from a fixed point below. Their contraction probably tends, therefore, to create an eversion of the mucous lining of the rectum during the act of defecation, and to bring the axis of the rectum more into the line of the anal canal.

The act of defecation is probably preceded by a marked relaxation of the levator ani and the sphincter muscles.

THE UTERUS.—Some points pertaining to the topographical relations of this organ have already been alluded to, in connection with the vagina and rectum; yet much remains that is not as yet positively settled, and which must continue, of necessity, to remain so, until the profession is sufficiently alive to the importance of the subject to undertake more extensive observations upon the living woman. Among those who have devoted special attention to the normal position of the uterus and its relations may be mentioned Schultze, Martin, Fritsch, Van de Warker, and Foster, each of whom has studied this subject upon the living female; while Panas, Aran, Scanzoni, Courty, Follin, Derneuil, Depaul, Goupil, and many others have contributed to the literature of this organ, especially as to the *attitude* which the organ assumes in the majority of subjects. If a comparison be made between the cuts found in most of the modern text-books, and the frozen sections of Braune, Heitzmann, Fürst, Hart, Kohlrausch, and many others,[1] it will be evident that the normal position and attitude of the uterus is still in dispute. Schultze describes in detail his method of making observations during life to determine

---

[1] Hart and Barbour have lately reproduced most of the best sections in their late work, "Manual of Gynecology," New York, 1881.

this point, which is certainly ingenious;[1] and Foster also publishes a drawing of an instrument made by him for the purpose of determining the normal direction of the uterine canal. The latter author (Foster), in addition, has devised a series of measurements which will enable an observer to place the os externum in its proper relations to the surrounding organs and pelvic outlines, in preparing a schematic drawing of any individual case. The monograph of Martin, on the contrary, fails to give the details of his method of observation; hence his deductions[2] are of less value than those of Schultze and Foster, as their accuracy cannot be verified properly without following the method employed by him, provided it is free from apparent sources of error.

The cuts found in the works of Gray and Savage, and many other anatomists, who have relied exclusively upon dissection as a means of ascertaining the normal position and attitude of the uterus, are so markedly in error (if the frozen sections of authors previously quoted be taken as a more reliable standard), that some of them are absolutely without value. If any reader desires to convince himself of this fact, let him compare the relative position of the tip of the coccyx with the plane of the lower border of the symphysis pubis in the cuts of these authors. The former point should lie at least one-half an inch higher than the latter in the standing posture. With a defect such as exists in the drawings of the anatomists named, how can a cut prove of value to a student of the normal topography of the pelvic viscera? If the cervix is placed in its proper relation to the lower border of the symphysis pubis in such a drawing, it is distorted in its relations to the sacrum, and vice versa. It

---

[1] Regarding Schultze's method, Dr. H. J. Garrigues, in a recent paper, comments as follows: "Such experiments would be absolutely impossible in this country. Our conscience, as well as fear of a suit for malpractice, and probably manslaughter, would restrain the physician, and not even the poorest woman in a charity hospital would submit to having a board pressed against her abdomen, sounds introduced into uterus and bladder, staves hooked to her womb, and being made to stand up and lie down at the command of the explorer. Besides, I do not think these methods are necessary. A guillotine, regulated by clockwork, is not needed for chopping wood."

[2] This paper was published posthumously. It is probable that it would have been more complete if the author had personally supervised its publication.

is certainly time to discard many of the cuts which are being perpetuated by incorporation in our popular text-books; and I would call the attention of authors to this defect which is so commonly repeated as to seriously impair the usefulness of their work. Even if a drawing be designed purely for a diagrammatic purpose, there is no excuse for the incorporation of so gross an error that may mislead the reader in his conception of surrounding parts—which the cut, perhaps, is not intended to show, but which it nevertheless represents incorrectly. The criticism made, however, is not upon cuts which are designated as diagrammatic, but upon those which are apparently intended to be actual representations of the parts, as revealed by dissection. In Savage's drawing of the sagittal section of the pelvis, not only are the bony parts distorted (as specified in a foot-note on a previous page), but the rectal walls are represented as nearly one-half the thickness of the uterus, the anterior vaginal wall below the urethra as even thicker than the womb itself, the anterior lip of the cervix as the longer, and the vagina as an open tube.

The objection which I would make to the method devised by Foster, and the same is true to a less extent of that employed by Schultze, is that the direction of the canal of the *cervix only* is determined, rather than that of both the cervix and the body of the uterus. In the cut which Foster publishes, as representing what his researches have led him to believe is correct, the canal of the cervix and that of the body of the uterus have the same direction; and it is the rule to find the same error made in almost all of the drawings, which have been incorporated in the later works upon gynecology.[1] I deem it to be an error, because the prevalent opinion which exists in the minds of those who have devoted the most time to the determination of the normal attitude of the uterus is decidedly in favor of a *slight degree of anteflexion*. In the exhaustive review of this subject by Panas, who himself examined 114 young or adult women, the opinions of Scanzoni, Boullard, Follin, Derneuil, Aran, Goupil, and others, are brought forward to support his statement that anteflexion must be con-

[1] Fritsch puts the angle formed by the cervix and body of the uterus at 90°. Schultze considers that the anterior surface of the uterus lies nearly parallel with the anterior wall of the vagina when the bladder is empty. Few observers coincide with these extreme views.

sidered the normal position of the organ in the unimpregnated, and nulliparous female. In his own researches, he found that the number of straight canals, as compared with that of the anteflexed womb, was about on an equality; but Goupil found that 41 out of 115 cases presented the condition of anteflexion and Aran two-thirds of the number examined by himself. In the total number of 333 cases, collected from reliable sources, Panas demonstrates that, while all attitudes of the uterus were at times discovered, the largest proportion of cases favored the view that anteflexion to a moderate degree must be considered the normal one; the unbent organ, which is considered by many as the only physiological one, being found in but one-third of the total number. He also concludes that the uterus probably tends to become straight with growth, after the age of puberty.

Regarding this deduction, I am aware that the following objections may be made: 1. That no post-mortem results are an infallible guide in establishing the position or posture of an organ which is capable of marked displacement during life. Some might even go so far as to assert that this method of research is absolutely without value, as a basis of deduction. 2. That normal uteri are uncommon, and that abnormalities in position and posture are more often observed than would seem, at a first glance, to exist in young subjects who have been free during life from apparent uterine disorders and the effects of pregnancy. 3. That conjoined or bimanual palpation was not practised when many of the earlier deductions as to the normal posture of the uterus were first given to the profession. 4. That the condition termed "flexion" is a misnomer if applied to very slight curvatures of the canal. 5. That no deductions as to the normal posture of the womb can be considered as reliable and worthy of credence unless all these elements of error can be shown to have been anticipated and guarded against by the investigator, and a sufficient number of experiments recorded to warrant positive conclusions.

In answer to these criticisms, a reply might be made (1) that post-mortem examinations of young nulliparous women, free from suspicion of existing uterine disease, is far more reliable as a means of determining the direction of the uterine canal than it might be for the solution of problems in topographical anatomy. The healthy organ is not so weak in structure as to bend of its own weight after death to an extent requisite to insure a permanent distortion after its removal from the pelvis. 2. That bimanual examination is not a reliable method to determine the point at issue, as the pressure required to bring the uterus within the sense of touch is liable to be a cause in producing its displacement. 3. That the methods employed during life by the observers mentioned (Schultze, Fritsch, Martin) are not altogether free from sources of error; and that the results obtained seem to demonstrate their fallacy, since few clinical observers accord with the deductions. 4. That post-mortem examinations are probably the most reliable means of determining the existence of morbid processes—excelling all the revelations given to us by the senses of touch or sight during life.

The view that anteflexion is the normal posture of the uterus is sustained, moreover, by most of the frozen sections of the female pelvis, which have been prepared with extreme care;[1] although, as has been stated in a previous page, these frozen sections are not an infallible guide in studying disputed points in the topography of the pelvic viscera. The extent of anteflexion found in some of these frozen sections is far greater than I believe to be normal; since it is enough to apparently impede the normal freedom of escape of the menstrual flow or the introduction of a uterine sound. Schultze concluded that the normal attitude of the uterus was that of anteversion, and the majority of authors seem to agree with him in that apparently erroneous statement. All authorities, however, concur in the opinion that the condition of the bladder modifies the posture of the uterus as well as the direction of the vaginal axis; and it should be said that the statements quoted above refer to the condition present when the bladder is empty. As the bladder becomes distended, the uterus is affected by the interposition of a tumor between it and the anterior wall of the pelvis; most authors attributing to it the power of creating a tendency toward a backward displacement of the uterus. An ingenious suggestion has been advanced by Joseph, in antagonism to all previously accepted ideas regarding this point, viz.,

[1] See the plates in Kohlrausch, Le Gendre, Pirogoff, Beigel, Simpson, Hart and others.

that as the bladder becomes distended, it requires more peritoneum to cover it than when empty, that it derives this excess of peritoneal covering from the anterior portion of the body of the uterus; and, that, above a certain level, the peritoneum refuses to leave the uterus, is rendered tense by the enlarging bladder, and actually becomes a tractor upon the uterus tending the pull it forward. My own investigations have convinced me that the attachments of the peritoneum to the uterus are more firm in front than upon the posterior surface of the organ; but I am not yet convinced that this theory of Joseph can be sustained by clinical investigation or experiment upon the cadaver, however ingenious it may appear, although the

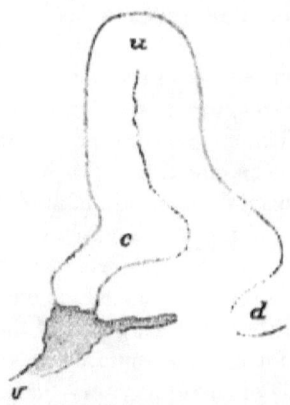

FIG. 8.—A sketch illustrative of the attitude of the virgin uterus, as revealed in some of the frozen sections of the pelvis (traced from Hart—Plate No. 1). It is evidently a distortion of the normal posture.
 u, Body of uterus; c, Cervix; v, Vagina; d, Cul-de-Sac of Douglas.

plate of Kohlrausch may possibly be brought forward in its support (Fig. 51 of Hart and Barbour. Wood's edition, New York, 1883).

Whether the bladder be full or empty, there is reason to believe that the body of the uterus is seldom thrown so far forward as to allow of the descent of the intestines into the pouch of Douglas when the woman stands erect.[1] If anteflexion be accepted as the normal posture of the uterus, the degree of curvature of the uterine canal must, to my mind, be considered

---

[1] Hart and Barbour dispute this statement, but acknowledge that it is sustained by most authorities (page 41, Wood's edition).

as much less than that depicted in most of the frozen sections. The beautiful plate of Kohlrausch, in which the bladder is represented as distended, was prepared from a frozen section with every possible precaution against error in depicting the normal topography of the pelvic viscera; it shows the uterus in a slightly anteflexed posture, thus apparently sustaining the deductions of Panas. The direction of the canal of the cervix is represented as nearly vertical, while that of the body of the uterus forms an angle of about thirty degrees with the vertical line. Copies of this plate are usually inaccurate.

A superior view of the pelvic viscera shows certain points in the topographical relations of the uterus which are not appreciated in an antero-posterior median section of the pelvis. It will be seen from this view that the uterus and its broad ligaments form a partition which extends transversely across the pelvic cavity; dividing it into a large anterior portion (which reveals the situation of the bladder, the round ligaments, and the utero-vesical ligaments), and a smaller posterior portion (which contains the rectum, the pouch of Douglas, and the sacro-uterine ligaments). The peritoneum, which is spread over the pelvic viscera like a sheet (dipping down between them and assisting to form the utero-vesical, broad, and sacro-uterine ligaments), often conceals the outline of the ovaries in looking into the pelvis from above. The diagrammatic representation of Hodge shows the general arrangement of the parts, as viewed from above, better than any with which I am acquainted. The broad ligaments are placed by him further forward than in most cuts of the pelvic brim, and I am inclined to believe, from my own observations, that they are correctly drawn. The investigations of my friend Professor Polk, who has studied this subject with great care upon a number of females of different ages, confirms the statement of Hodge that the attachment of the broad ligaments to the wall of the pelvis is situated in the nulliparous and unimpregnated female along a vertical line which extends between the great sciatic notch and the obturator foramen, and not as far back as the sacro-iliac synchondrosis, as is stated by some authors. The investigations of Polk seem also to demonstrate that the point of attachment of these ligaments changes during the pregnant state, and more nearly approaches the sacro-iliac synchondrosis at full term.[1] The peritoneum is seen, in this diagrammatic cut of Hodge's, as a thin sheet which is tucked (to use a homely expression) between the rectum, uterus, and bladder, and whose folds are prominent in the region of the various ligaments connected with the uterus. A second diagram by the same author, representing a posterior view of the uterus and its ligaments, may well be studied in connection with that of the superior view; since both will help to convey a more complete conception of

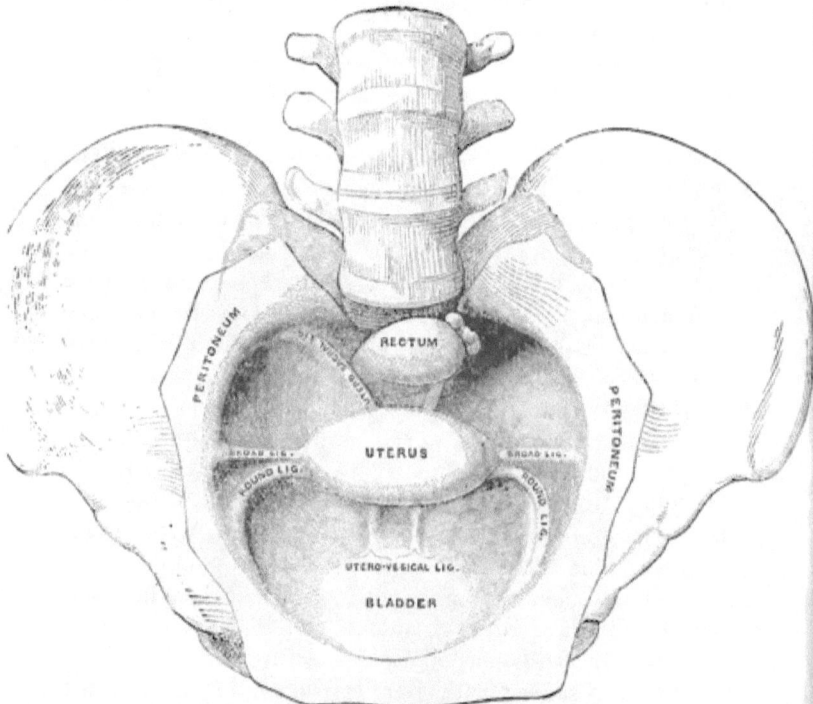

Fig. 9.—A diagrammatic superior view of the female pelvis, showing the reflections of the pelvic peritoneum (Hodge).

the relations of the uterus to the reflections of the peritoneum and the rectum. This drawing shows the Fallopian tubes and the ovaries lying in relation with the uterus and the broad ligaments; and the pouch of Douglas (the lowest point in the peritoneal cavity), extending between the uterus and the rec-

[1] Hart and Barbour state that the peritoneum is stripped off from the bladder during pregnancy. The broad ligaments become nearly vertical at full term.

tum, and reaching down below the point of attachment of the vagina to the uterus.[1] This reflection of the peritoneum can be studied best, as regards its relations to the uterus, vagina, and rectum, in the pictorial representations of antero-posterior median sections of the pelvis given in previous pages. The sacro-uterine ligaments which have been referred to as forming a part of the upper segment of the pelvic floor are also shown in this diagram, diverging to reach their sacral attachments; the intervening space being filled by the rectum, as is ap-

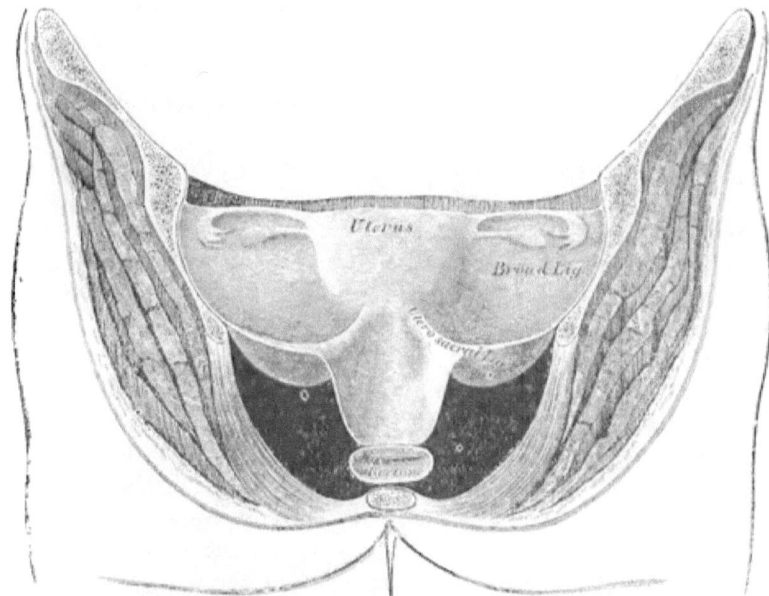

FIG. 10.—A partly-diagrammatic transverse vertical section of the female pelvis, showing the relative situation of the peritoneal and subperitoneal pelvic spaces, and the pouch of Douglas (Hodge). The ovaries and Fallopian tubes are not represented in accordance with the views of the author.

parent in Figs. 7 and 9. Finally, this diagram shows that the peritoneum of the pelvis does not extend downward as far as the level of the pelvic floor, a space (darkly shaded in the cut) being left, called the "sub-peritoneal" space, in which are situated the lower portion of the rectum, the vagina, a part of the uterus, the ovaries, the bladder, urethra, and ureters, and

[1] This pouch varies in its depth. It may extend 3 cm. (1¼ inches) upon the posterior wall of the vagina. It usually covers only 12 mm. of the vagina. In one frozen section, it extended almost to the extreme limits of the vagina (Pirogoff).

a mass of cellular tissue, blood-vessels, lymphatics, nerves, and elastic and muscular fibres. In fact, all of the pelvic organs are placed underneath the pelvic peritoneum; although the fundus of the uterus rises above the plane of the bladder when that organ is empty, and thus appears to be more completely enveloped by the peritoneum than the contiguous bladder, or the rectum which lies behind it. The elasticity of this peritoneal covering seems to be enormous. In spite of the immense size that the uterus, bladder, and ovaries sometimes attain, the attachments of the peritoneum are but slightly disturbed in proportion to its great distention. The experiment of Savage, who tied the urethra and injected the bladder through the ureters in order to simulate the natural method of its expansion, demonstrated to his mind that the pubo-vesical pouch of peritoneum was not appreciably raised. This elasticity is evidenced to a still greater degree by the fact (which seems to be well proven, both by clinical observations and frozen sections of the pelvis), viz., that the pouch of Douglas is not raised to any marked extent even at the full term of pregnancy.

The fundus of the normal uterus does not usually extend much above the level of the plane of the brim of the pelvis in the corpse; although the entire organ[1] surmounts that plane, which passes horizontally from the upper border of the symphysis pubis. I am inclined to believe that the deduction of Sappey that the fundus normally lies three-quarters of an inch below the plane of the pelvic brim is not true in the living subject. My researches convince me that the fundus should be placed slightly above that plane, in any schematic drawing of the pelvic viscera in an antero-posterior median section.

The non-pregnant, pregnant, and even the fetal uterus shows a normal inclination toward the right side, and the same deflection has been proved to exist in the uterus after the pregnant state by Pfannkuch. This is not usually shown in the cuts made to represent the view of the pelvis from above.

Now, let us compare the cut which I offer to the profession with one devised by Thomas (Fig. 43 of "Dis. of Women")

[1] This statement I believe to be correct. Almost every section of the frozen subject will sustain this view, as can be tested by any reader who is in doubt concerning it.

which he considers to be an accurate conception of these parts. We shall see in the latter that the tip of the coccyx is so much out of its normal relation to the symphysis as to lie apparently

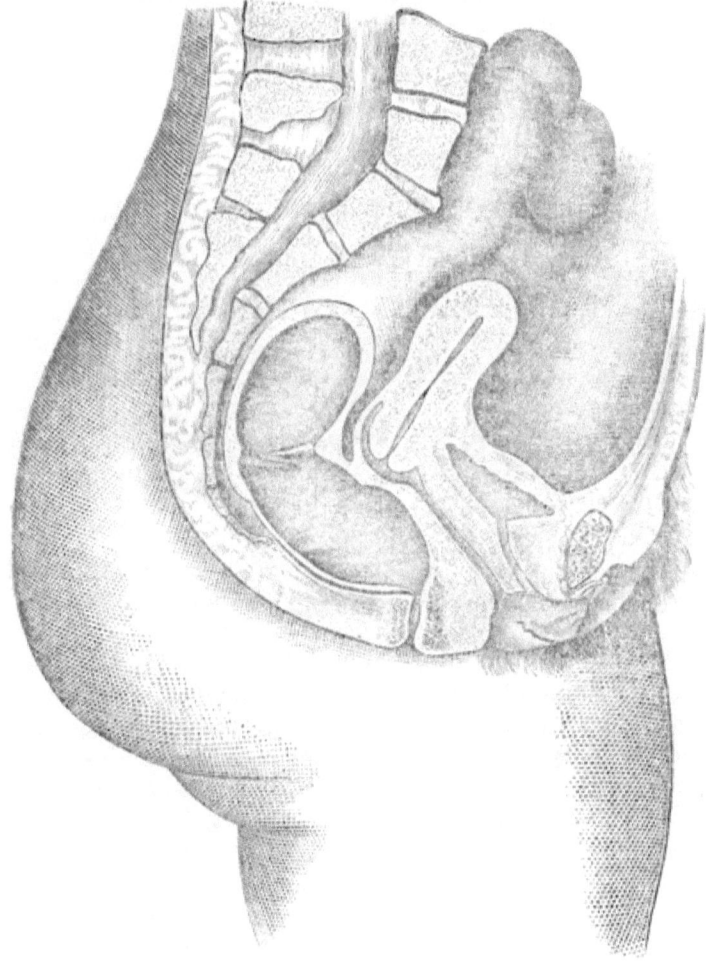

FIG. 11.—The author's schematic drawing of the sagittal section of the female pelvis and its organs.
The uterus is not flexed in the cut quite to the extent shown in the author's original drawing. The degree of flexion shown in Fig. 1 is about correct, according to his views.

some inches below its proper level; that the cervix touches the horizontal plane passing through the lower border of the symphysis; that the internal orifice of the urethra appears to be much below its proper level; that the rectum appears as a round tube cut open, rather than as a flat section of a collapsed canal; that the anal canal (one inch long) is absent; that the uterus is straight; and that the meatus urinarius is at least twice as far from the symphysis as it should be. Again, if contrasted with the drawing of Savage ("Female Pelvic Organs," plate XIII. of Wood's edition), the same error in regard to the relative position of the tip of the coccyx to the symphysis exists; the sacral canal is invested for its entire length by bone; the symphysis is excessively deep; the urethral meatus lies far below the level of the symphysis; the ostium vaginæ is below the plane of the meatus urinarius; the anal canal is absent; the anus is patulous, and the rectum incorrectly shaded; the anterior vaginal wall is thicker than the uterus; the entire bladder (moderately distended) lies below the plane of the upper border of the symphysis; the uterus is intersected by the same plane in the region of its fundus, when it should pass below the level of the cervix; the curve of the nates is erroneous; the angle made by the plane of the superior strait of the pelvis with the vertebral column is incorrect; finally, the plane of the inferior strait shows the point of greatest projection of the soft tissues below it to be at the middle of the labium rather than the region of the anus.

These criticisms may appear to be too positively stated, but they are sustained, I think, by a close analysis of most of the frozen sections hitherto published, as well as by careful measurements made upon the living woman. They are due, to a great extent, to the fact that the artist has distorted the relations of important points in attempting to draw an upright figure from a subject lying upon the back. This, of course, is not the case in a photograph of a frozen section.

Points of special interest pertaining to the Uterus.—The round ligaments of the uterus have been studied in detail by Rainey, who describes them as arising by three tendinous fasciculi—the inner being connected with the tendinous expansions of the internal oblique and transversalis muscles of the abdomen, the middle from the external abdominal ring, and the outer from the neighborhood of Gimbernat's ligament. These three fasciculi unite to form a rounded cord, which passes through the inguinal canal in front of the deep epigastric artery, and then downward and inward between the layers of the broad ligament of the uterus to reach the anterior and upper part of the organ. Some points pertaining to the physiology of these ligaments will be discussed in subsequent pages.

The base of the broad ligament can be reached by the finger when it is crowded against the lateral wall of the fornix vaginæ. This point may be utilized in diagnosis. These ligaments are altered in their relations to other pelvis structures during gestation. This will be discussed later.

There is a large amount of tissue between the layers of the broad ligaments, which is rich in blood-vessels. Sections of the pelvis in the antero-posterior plane show that this tissue is most abundant where the broad ligaments join the uterus, and that it decreases steadily as the pelvic wall is approached.

Horizontal sections, as made by Pirogoff and Ruedinger, show the existence of a loose fatless tissue, abundantly supplied with blood-vessels and lymphatics, which invests the lower part of the uterus and the upper part of the vagina. This is the "parametric tissue" of Virchow, which has been described also by Spiegelberg. It averages about two centimetres in thickness. Ruedinger's plate is reproduced by Hart and Barbour, where it can be studied by the reader.

Clinical observation has led almost all gynecologists to accept the gross inaccuracy of a retroverted uterus as its normal posture, in spite of the drawings of Luschka, Cruveilhier, Henle, Braune, and others. To what extent the uterus tilts forward during life, it is perhaps impossible to ever decide positively, as the method employed by Schultze would appear to antevert the organ unduly and bimanual palpation would also tend to do the same. All frozen, spirit-hardened, and chromic acid preparations have sources of error which cannot be eliminated. These have been pointed out in previous pages. Foster's method only determines the direction of the axis of the cervix. The subject has, however, been discussed in preceding pages.

The uterus, when normally situated, and the bladder move together, according as the varying degrees of distention of the latter compel movement.

In the genu-pectoral posture, the relations of the uterus,

vagina, bladder and pelvic peritoneum are materially modified.

By digital pressure the uterus can be raised about one inch and a half. It can be drawn downward by the volsella to the ostium vaginæ without endangering its return to its proper position in the pelvis. It is extremely mobile in the antero-posterior and lateral planes—its ligaments not being tense in any direction.

The *volsella* enables the gynecologist to make an accurate diagnosis of all abnormal conditions of the cervix; it increases the scope of rectal exploration; it is of use in almost every form of surgical procedure upon the uterus; it helps to establish suspected attachments of the uterus to tumors of the abdomen; it aids in the replacement of the gravid or non-gravid retroverted uterus; it simplifies the introduction of tents; finally, it is indispensable in operations for the repair of lacerations of the cervix, amputation of the vaginal portion of the cervix, and the removal of the uterus through the vagina for malignant disease.

Many points pertaining to the ligaments of the uterus will be found incorporated in those pages which treat of the pelvic peritoneum, the pelvic cellular tissue, and the ovaries.

If the broad ligament of the uterus be held up to the light, the parovarium or organ of Rosenmüller may be seen between the ovary and the ampulla of the Fallopian tube. These bodies are occasionally the seat of degeneration, producing the cystic tumors of the broad ligament known as parovarian.

The course of the Fallopian tube within the cavity of the broad ligament is commonly compared to that of a shepherd's crook. From my own dissections, I am inclined to regard its course as comparable to a "surcingle" to the ovary. This view is supported by some of the later observations of Tait, Doran, and others. When positively settled, it will have a tendency to shed much light upon the diagnosis of tubal cysts and tubal pregnancy in their early stages.

The structure of the mucous lining of the uterus (although in no way connected with its topographical relations) opens up fields for investigation, respecting its glands, epithelium, its menstrual changes, and its functions during gestation.

It is important to note that horizontal sections of the pelvis show marked variations in the uterus, since the shape of that organ changes with the altitude of the section. Farre has given diagrams to illustrate this point, in his contribution to the Cyclopedia of Anatomy, which have been incorporated by Hart and Barbour in their late work.

The relative position of the uterus, and its annexa, to adjacent parts in the neighborhood of the pelvic brim, has been studied, through their peritoneal covering, by Hasse of Breslau, who froze a subject and then lifted out with great care the soft viscera without disturbing the pelvic contents. The drawing which he was thus enabled to prepare has been copied by Hart and Barbour. A coronal section of Ruedinger's, and a drawing of Schultze's have also been utilized by the same authors to show the various views of the same.

The extreme mobility of the uterus allows of almost incessant change in position during life. Among the more important factors which tend to produce such changes, may be mentioned the acts of respiration, singing, and walking, and all violent muscular efforts, in addition to the mechanical effects of rectal and vesical distention.

Guerin asserts that the cellular tissue of the broad ligaments of the uterus can be demonstrated by means of inflation to be perfectly separate from and independent of all association with that of other parts of the pelvis—a fact which other observers deny. The same author asserts that there is no proof of any special inflammatory affection of these ligaments which can be diagnosed during life.

The sacro-uterine ligaments are rendered prominent as distinct cords (which can be felt through the rectal walls) if the uterus be drawn downward by the volsella.

The vessels of these ligaments may cause troublesome hemorrhage, if injured.

Pregnancy causes an alteration in the height of their sacral attachments. They are attached nearly on a level with the promontory of the sacrum at the thirty-sixth week of gestation (Polk).

The alteration in the position of the pelvic peritoneum during the pregnant state causes the broad ligaments of the uterus to become markedly changed, in respect to their shape and boundaries, from their normal condition. The upper border of each becomes nearly vertical; the base rises to the level of the pectineal line and becomes greatly increased in its antero-posterior measurement (extending from the pectineal eminence to the sacro-iliac synchondrosis, on account of its reflections from the round ligament and the ovarian artery); finally, its shape becomes triangular, as the result of the widening of its base.

The broad ligament may, and may not, contain the ureter between its laminæ in multiparous subjects. In the virgin and nullipara, the ureter does not lie between its layers (Polk). These variations are due to the tendency of the broad ligaments toward a backward displacement during gestation.

THE BLADDER.—The topography of this organ has already been discussed in connection with its influence upon the posture of the uterus, and its physiological connection with the anterior wall of the vagina. These points need not be repeated, as they can be found by referring to previous pages.

The *shape* of the bladder, when collapsed, is differently represented. Henle, in a sagittal section of the pelvis, shows the bladder to be round, small in size and hence probably contracted, and with a distinct cavity which also presents a rounded outline. Braune, in a frozen section, depicts the organ as so completely collapsed as to present scarcely any apparent cavity, and with an outline which is a decided ovoid; but in two woodcuts of sections made by Le Gendre, incorporated in Braune's work, the organ is again represented as round, although apparently collapsed, and as possessing a cavity of nearly a corresponding shape. Pirogoff, in his section made through the antero-posterior diameter of the pelvis in the median line, found a collapsed bladder which was markedly triangular in form and which presented no cavity. Most of the woodcuts in the popular text-books represent the bladder as distended to a greater or less extent; hence they are of no value in determining the condition which is to be considered as normal to the collapsed organ, even if their accuracy as guides in many other particulars could be relied upon. The wide variations in the appearance of this organ, in those sections of frozen subjects which are presumed to depict it as empty, attracted the attention of Braune; who states that he considers the rounded form as the normal one,[1] and the triangular outline to be the result of decomposition in a subject, who had passed urine just before death and thus caused the walls of the organ to be unsupported. He denies the statement of Claudins, that the uterus is not affected in its posture by the distention of the bladder, and asserts that changes in the attitude of the uterus may be easily detected during life, as the act of micturition is being performed.

Now, as I am inclined to differ from so high an authority as to the normal configuration of this organ when collapsed, I will endeavor to adduce reasons which seem to me to sustain the triangular outline rather than the round. We know in the first place that the bladder is intimately connected to the anterior wall of the vagina (in its lower part), and that cellular tissue and the peritoneum bind it to the anterior wall of the uterus. We also know that the peritoneum is so closely adherent to the anterior surface of the uterus[2] as to be undetachable, and we have apparent reason to believe that slight anteflexion of that organ may be considered its normal posture. In addition to these facts, we are in posession of most positive knowledge that the intestines rest upon the upper part of the bladder and are capable of exerting a downward pressure upon it. Now, from these three facts, the most plausible view would seem to be that the anterior vaginal wall—connected, as it is, to the sacrum indirectly by the uterus and the sacro-uterine ligaments—would tend to act to some extent upon the contiguous bladder as a line of tension; that the uterus, which is displaced by the accumulation of urine, would naturally tend to compress the bladder, as it became emptied, on account of the recoil of the elastic peritoneal covering which is continued from its anterior surface upon the bladder and from it to the abdominal wall; and, finally, that the weight of the intestines above would be in excess of the normal tonicity of the vesical walls, and would therefore assist in producing a close approximation of these walls when the urine had been expelled from the cavity of the organ.

I am inclined to believe that the entrance of the urine into the

[1] Savage also distinctly states this to be his opinion. The frozen section of Heitzmann shows a rounded outline does sometimes exist.

[2] Savage states that the same is true of the peritoneal reflection from the bladder to the anterior wall of the abdomen. This is denied by Hart.

cavity of the bladder is assisted to some extent by the "piston-like" action of the diaphragm,[1] which tends to draw the abdominal organs upward during expiration; because its curve becomes greater at that time, and its convex surface projects far into the cavity of the thorax. "Nature abhors a vacuum"—and, as the air is expelled from the lungs, the tendency of all the movable viscera of the abdomen is to rise to fill the space thus made vacant. I do not mean to discard the effect of the downward pressure of the urine in the ureters and renal pelvis and the blood-pressure in the vessels of the kidney as a great force in distending this organ in spite of the weight of the viscera which tend to compress it; but I think the pressure of the intestines is greater during inspiration than when air is being exhaled, and that the bladder is in a condition which particularly favors the entrance of urine, during expiration. The triangular appearance of the bladder when collapsed is moreover sustained by the sagittal sections of the pelvis made by late authors, among whom may be mentioned Fürst, Schultze, Fritsch, Pirogoff, Waldeyer, Hart, and Beigel; and the condition found in the plates of these authors will be described as an approach to that which, to my mind, is the normal appearance of this organ when completely emptied.

Hart, of Edinburgh, who has lately come into prominence as an original investigator in female pelvic anatomy, states in recent article that he is inlined to regard the bladder as an organ which is capable of both a *systolic* and *diastolic* action. The former (systole) assists in the expulsion of the accumulated urine from its cavity, aid being likewise furnished by the pressure of the intestines which is exerted at a right angle to the peritoneal surface of the organ. The latter condition (diastole) is brought about in order to aid in the escape of urine from the ureters, the muscular walls exhibiting a state of extreme relaxation. He supports the view that sagittal sections of the bladder reveal the triangular form in many instances, although he also believes that the anterior and posterior walls of the organ may be brought in contact and thus produce a form which is convex upon its peritoneal surface. He states that the method

---

[1] The reader is referred to Duncan, Busey, Taylor, Thomas, and the lectures of Prof. Küss for a more detailed statement of this action of the diaphragm.

of folding of the walls of the collapsed bladder produces two distinct types of outline of the vesical mucous surface: 1. A curved slit (continuous with the axis of the urethra), when the walls collapse from before backward. 2. An outline which corresponds to the letter Y (the posterior limb being sometimes the shorter), when the upper wall is approximated to the lower.

From our present knowledge, it is impossible to state positively which form of the collapsed bladder is to be considered as the normal type. It is possible and perhaps probable that more than one may exist, and that each bladder may have an individuality in the method of folding of its walls. I am glad, however, to bring the views of Hart before the profession, as confirmatory of my own position, in regard to the triangular outline of the collapsed organ; since the view of Braune would lead to the inference that all organs found with the upper wall in contact with the lower were altered by post-mortem changes.

The base of the vesical triangle (see Fig. 11) seems to correspond with the peritoneal surface of the organ, which presents a direction nearly horizontal. The sides seem to be formed by the folding of the anterior and posterior surfaces of the organ upon themselves. The apex appears to correspond to the situation of the internal opening of the urethra. This is perhaps not in accordance with the usual description,[1] as the appearances differ in almost every sagittal section of the pelvis, but it is sustained by the plates of the authors, quoted as exhibiting a triangular section of the collapsed organ. The flattening of the peritoneal surface of the bladder, which presents in some sections even an indented appearance) would certainly appear to sustain my view of the importance of intestinal pressure as a factor in creating this mode of collapse.

The bladder is bound to the upper part of the anterior wall of the vagina by loose connective tissue;[2] the surface of the

[1] Some authors describe the situation of the base and apex of this triangle differently. But, as the sides of the triangle appear nearly equal in length, I have selected, as the base, that side which appeared to me to be longest in the majority of frozen sections. When the Y-shaped bladder (diastolic organ of Hart) is found, the base of the triangle has become indented.

[2] In the upper part of the vagina, the vesico-uterine pouch of peritoneum approximates is so closely as to render operative procedures in this locality liable to a serious complication.

organ, which is so attached, having (as Garrigues described it) the shape of a heart. The boundaries of this heart-shaped space are as follows: the anterior or lower limit runs parallel with, and slightly external to, the boundary line of the "trigonum vesicale;" the upper limit follows the outline of the limits of vagina, and crosses over the cervix of the uterus. It is in this region that the ureters have the most intimate relation to the bladder and vagina. The distance from the internal opening

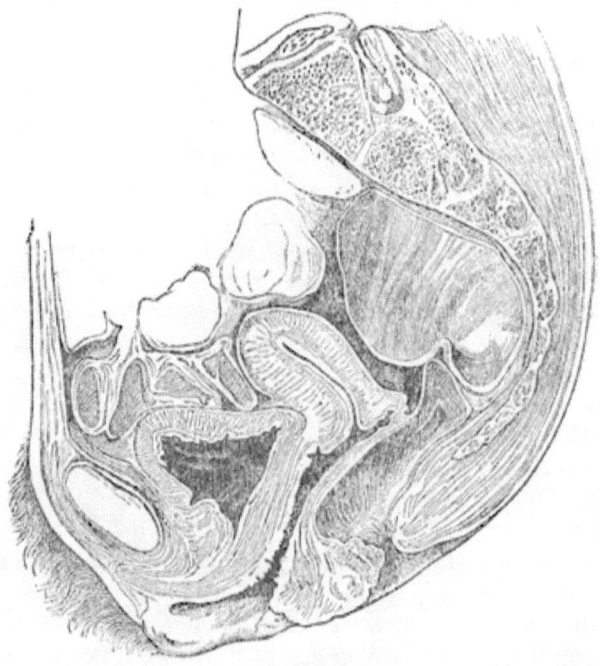

Fig. 12.—Sagittal frozen section of the female pelvis (Waldeyer). This section appears to justify the view that relaxation of the anterior wall of the vagina has allowed the bladder to sag, thus distorting the relations of the urethral openings to the symphysis pubis. It sustains the view, however, that the bladder collapses in a triangular form.

of the urethra to the cervix uteri has been carefully measured by Garrigues, who states it as 3.2 cm. (one inch and a quarter); while the same authority places the extent to which the bladder overlaps the cervix uteri at 1.5 cm. (about five-eighths of an inch).

The bladder is normally smaller in its vertical diameter in

of the Female Pelvic Organs. 59

the female than in the male, but is wider. In children it is conical, and points higher in the abdomen when moderately

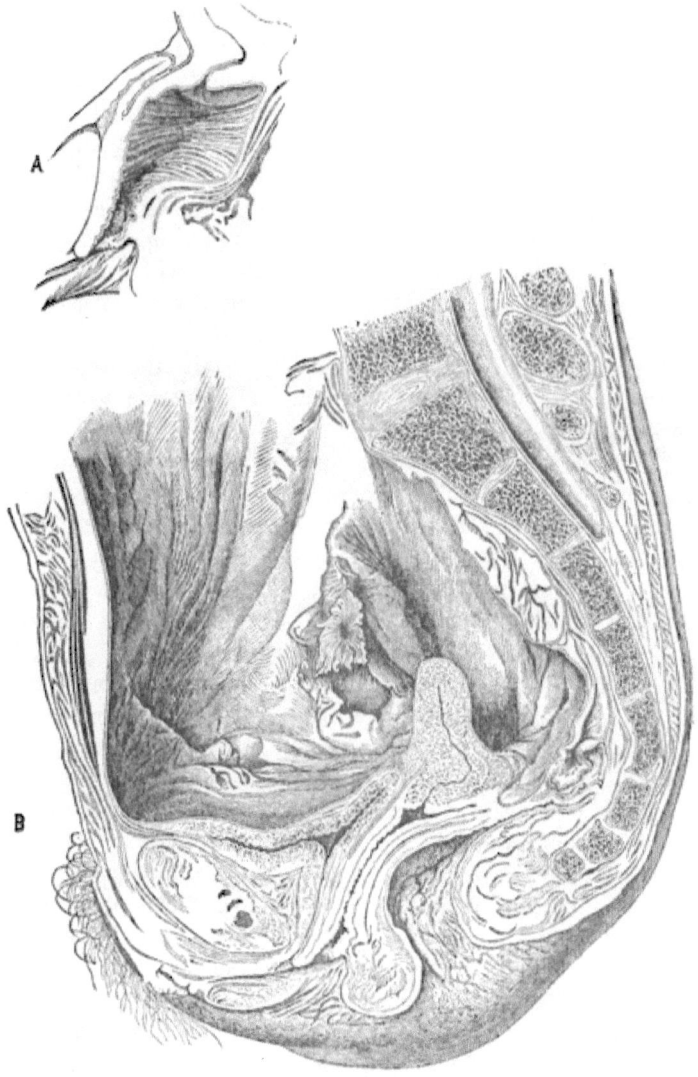

Fig. 13.—Sagittal frozen section of the female pelvis (Hart). A, fornix of the vagina and the cervix uteri, enlarged from B. Note the shape of the collapsed bladder; the form of the perineal body; the extreme flexion of the uterus; and finally, the relations of the bladder, uterus, and urethral openings to the several planes of the pelvis.

distended than in the adult, as the organ lies above the plane of the pelvic brim. In the adult, the bladder, when excessively distended, may reach almost to the level of the umbilicus; but, when collapsed, it lies normally below the plane of the brim of the pelvis, and completely behind the pubes. The collapsed bladder may sometimes be felt in the living subject by conjoined manipulation. The attachment of the organ to the pubes by areolar tissue is less firm than that between it and the uterus and the upper part of the vagina.[1] The capacity of the bladder is usually considered to be greater in the female than in the male. The relation of the internal opening of the urethra to the points of entrance of the ureters will be discussed later. The ureters will be discussed separately in subsequent pages.

Points of special interest pertaining to the bladder.—As a summary of previous pages, I would call attention to the following points.

The bladder when empty lies in front of and above the vagina and entirely behind the pubic bone. Its shape, when empty, varies with the condition of the organ, in respect to the tonicity of its walls.

It has three openings; the internal orifice of the urethra and the openings of the ureters. These openings are the landmarks for the division of the organ into three parts, as follows: the body, which includes all above the line which joins the internal openings of the ureters with the centre of the symphysis pubis; the base, which includes all below that plane; the trigone, which comprises that portion bounded by the internal orifice of the urethra and the line connecting the ureteric openings.

A valve-like action is attributed by some authors to a puckering of the mucous membrane in the region of the neck. It is still a matter of dispute among observers whether the bladder possesses a true sphincter muscle at its neck.

A solid ridge may be detected beneath the mucous coat of the bladder, which connects the two ureters. It was first described by Juric as the "inter-ureteric ligament."

The alterations in shape of the contracted bladder (as re-

---

[1] The connective-tissue behind the pubes (*retro-pubic fat*) has a *triangular outline* in all sagittal pelvic sections.

vealed by frozen sections) are to be explained by a systolic and diastolic vesical period. This point has been previously discussed.

In the parturient female, the bladder is drawn above the pubis; and the same displacement may be observed in case the gravid uterus becomes retroverted, since the organ is tilted upward.

Most of the ligamentous attachments of the bladder are formed by the peritoneum. The true ligaments of the organ are derived from the pelvic fascia.

As the bladder becomes distended, the uterus is crowded backward and the utero-rectal pouch of peritoneum is diminished in its antero-posterior measurement. The intestines, which gravitate normally to about the level of the utero-sacral ligaments when the uterus falls forward, are thus forced out of the utero-rectal pouch by the distention of the bladder.

Undue distention of the bladder is considered by some as a possible cause of permanent retroversion of the uterus, especially if that organ be gravid.

The height of the reflection of the peritoneum from the bladder upon the anterior wall of the abdomen is considerably increased by the full distention of that organ. This occurs also (although to a less extent) during advancing gestation, irrespective of the capacity of the bladder, on account of the traction exerted upon the organ as a whole. These points are disputed by Savage. Hart and Barbour sustain them, however, without qualification; and bring frozen sections forward in support of their views.

The *side-lateral* and the *semiprone* posture are best adapted for operations upon the anterior-vaginal wall. The former is often used in passing a sound or catheter; the latter is usually employed in operations for vesico-vaginal fistula. Occasionally, the *lithotomy* position is employed for operations upon the anterior vaginal wall or the bladder.

The reflection of the peritoneum from the anterior wall of the abdomen upon the empty bladder is about one inch and a half above the symphysis pubis in the genu-pectoral posture. In the erect posture it leaves the abdominal wall at the level of the symphysis.

The angle formed between the urethra and the empty bladder is affected by posture. In the upright attitude, the angle is nearly 90°; in the knee-chest position, no appreciable angle is formed, as they practically from a straight line (180°).

THE URETHRA.—In the lower or anterior portion of the upper wall of the vagina, the vesical structures and the tissues of the urethra are so intimately blended as to be practically inseparable. Blum, who has written one of the best monographs upon the female urethra, found the length of that canal to average about 3 cm.[1] He places the superior orifice or "vesical neck" as behind and slightly higher than the top of the pubic arch. This appears to the author to be an error. The external orifice is placed by him "at about 2 cm. behind the clitoris, a little in front of the tubercle which marks the end of the 'anterior column of the vagina.'" It is well to remember, however, that the situation of the meatus urinarius varies with the condition of the uterus and bladder, although these variations are not so marked as to possess any great importance. The urethra is separated from the vagina, in its upper part, by a space filled with cellular tissue and numerous veins; and similar structures exist between it and the sub-pubic ligament. It also bears relation with the anterior ligaments of the bladder and the roots of the clitoris. It is connected with the constrictor vaginæ (a part of the levator ani), the compressor urethræ, and the bulbo-cavernosus muscles. The canal is surrounded with a sheath of erectile tissue (*bulb of the urethra*), whose vessels are continuous with those of the vaginal bulbs.

The course of the urethra is given by most authors as straight; and corresponding nearly to a vertical line in the erect posture and to a horizontal line when the woman lies upon the back. Hart tates that its direction is parallel with the plane of the pelvic brim. I am inclined to believe that this is an error. The urethra seems to me to be curved, its concavity looking toward the symphysis pubis whose posterior and inferior portions it encircles. This view is supported by some late investigators, among whom may be mentioned Blum ; who states that the distance between the concavity of the urethral curve and the symphysis varies from 8 to 9 mm. The probable cause of this error

---

[1] Hart gives it as 1¾ inches, and its direction as parallel to the pelvic brim.

in description rests in the fact that the internal orifice is displaced, by anything affecting the posture of the uterus to an extent sufficient to pull upon the bladder or vagina, as well as by the simple distention of the bladder itself. It is not to be wondered at, therefore, that Hybord and other investigators, who appear to have failed to properly appreciate this fact, could not state the exact curvature of this canal.

The form of the meatus urinarius varies. Sometimes it is lengthened into an antero-posterior slit; again it may be asterated, or star-like in appearance; finally, it may appear perfectly round.[1] Upon either side of this opening, may be detected a smaller opening (the orifice of one of the so-called *glandulæ vestibulæ minores*) which may be mistaken for the meatus by beginners, in attempting to introduce a catheter. These are two of the small glands of the vestibule, and they may often be employed as a guide to the meatus which is placed between them. Some measurements, possessing a special surgical interest (into which the internal orifice of the urethra enters) will be found in those pages which treat of the normal course and topographical relations of the ureters.

Recently Skene, of Brooklyn, has described two tubular pouches in the female urethra, "lying near to its floor and extending up from the meatus for about three-fourths inch." They lie beneath the mucous membrane, within the muscular wall of the canal.

The question of the existence of true "sphincter muscles" to the urethra is not as yet positively settled.

Special Points of Interest Pertaining to the Urethra.—The sphincter muscle of the urethra seems to be partly an erectile structure. One of the consequences of this is that the conditions of the urethra due to laceration or excessive dilatation, although often recovered from, are liable to leave an incurable form of incontinence of urine.

The operation of *lithotrity* in the female is more difficult than in the male. This fact is chiefly to be attributed to the poor retentive power of the female organ, which allows the walls of

---

[1] According to Henle, the urethra appears as a sagittal slit on transverse section near to the meatus; a transverse slit near to the bladder; and star-shaped between the two points.

the bladder to become so closely applied to the blades of the instrument as to impair its working capacity.

Lithotomy may be performed either through the vagina, the vestibule, or the perineum. If vesico-vaginal lithotomy be performed, the fact that the vesico-vaginal septum becomes diminished in size as the bladder contracts and that the openings of the ureters are to be avoided should impress every surgeon with the dangers of ever making the incision a transverse one. If the vestibular method be thought of, it is important to remember that a calculus of even moderate size will cause laceration of the urethro-vesical sphincters, and that the knife must pass through the *urethro-pubic space*, which warrants serious objections to the operation. The operation of lateral lithotomy (as practised chiefly by Buchanan, of Glasgow) seems to allow of the extraction of calculi of the largest size with few if any serious obstacles. The incision is made in the delicate covering of the inner side of the left labium and the urethra subsequently opened, upon a sound passed within it, where it ceases to be incorporated in the anterior wall of the vagina.

The relations of the ureters to the vesical neck and also to the cervix uteri have important bearings upon the operations for vesico-vaginal fistula, lithotomy, and the removal of the uterus. These tubes will be discussed later.

THE EXTERNAL GENITAL ORGANS.—Something has been said already concerning the extent of projection of the soft parts beyond the plane of the pelvic outlet, and the relative distances of the fourchette, meatus urinarius, and clitoris, from the tip of the coccyx and anus. The vagina and urethra have also been considered in reference to their topographical relations; so that there remains only the component parts of the vulva to be discussed. In the literature of the ancients, as first pointed out by Goodell, the terms *vulva*, *matrix*, and *uterus* are used to designate the genital canal as a whole. The former term is now employed, however, to designate that portion only which lies external to the hymen, which marks the seat of the external vaginal orifice. The researches of Budin, which have been referred to on a previous page, have thrown light upon the construction of this membrane and have proved that it must no longer be considered as a simple reduplication of the mucous membrane, but as a *continuation* of *the vagina as a whole*.

In treating of the topography of the vulva, the fourchette labia majora, and labia minora (nymphæ) particularly invite a description.

The *fourchette* consists of a fold of skin which bears a different relation to the surrounding parts when the labia are in contact or separated. When the woman is in the erect, as well as in the supine position, the labia are closely approximated; hence the fourchette bags downward as a loose fold, stretching from one labium to the other, in the region of the so called posterior commissure of the vulva. If we separate the labia, how-

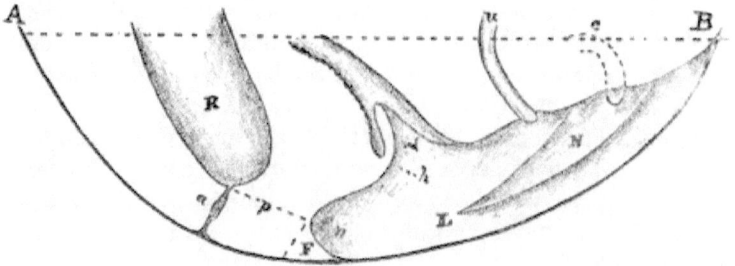

FIG. 14.—A diagram intended to show the relation of the pelvic floor and external genitals to the horizontal plane, in the erect attitude of the female (modified from Garrigues). A-B, horizontal plane; C, clitoris appearing as a bulb between the nymphæ (N); *u*, the urethra, opening posterior to the nymphæ (this canal is curved during life); V, vaginal orifice pulled somewhat open to show the most common posture and seat of the hymen (*h*) when the vagina is closed; R, rectum narrowing into the anal canal (*a*) as it perforates the pelvic floor; L, labia majora; N, labia minora, or nymphæ; F, fourchette, the dotted line showing its position when the labia are in contact, and the dark line the position when the labia are separated; *f n*, fossa navicularis formed by fourchette, when the labia are separated; *p*, line of pelvic floor between anus and rectum; *a*, anal canal; *h*, hymen, when vagina is closed.

ever, it will be seen that the fourchette becomes tense, and gradually advances until its anterior margin reaches the posterior border of the external vaginal orifice. The hollow which exists behind the fourchette, so stretched, is called the "fossa navicularis," from its fancied resemblance to the form of a boat. When the labia are in close approximation, this hollow is wanting. The surface of the fourchette which looks toward the vaginal orifice is red in color, moist, and marks the transition which here exists between the skin on its anterior surface and the mucous covering at the base of the inner surface of the nymphæ. This interior surface of the fourchette is continuous with the covering of the interior surface of the labia minora; which, since it possesses minute hairs, is considered as properly belonging to integumentary structures. The boundaries of the fourchette are as follows: above, by its free edge, when put on the stretch by separating the labia majora; below by the attachments of the integumentary fold to the labia and the perineal body. The fossa navicularis has different boundaries. Its upper wall corresponds to the perineal body, continued as far forward as the *ostium vaginæ;* its lower wall is formed by the interior surface of the fourchette, when the labia are separated (thus putting it upon the stretch). The anterior limit of the lower wall corresponds to the free edge of the fourchette. These facts will enable the reader to appreciate the bearing of a criticism of Garrigues upon the prevalent misuse of the terms, which I quote as follows: "when, therefore, a recent author advises to perform episiotomy midway between the fossa navicularis and the fourchette, his advice is as difficult to follow as to make an incision between the lower lip and the mouth." Hart and Barbour make a curious statement, which I deem to be an error, when they say, "When the fourchette is *pulled down* by the finger, a boat-shaped cavity is made—the fossa navicularis."

The diagram, which I have introduced as tending to make clear some of the points in the topography of the vulva, will render the special consideration of the topographical relations of the clitoris unnecessary. The pages which treat of the female perineum will cover any important omissions.

In the nude erect female, only the mons Veneris is seen, the external genitals being hidden by the thighs. The labia majora usually lie in close apposition, and are only separated to a slight degree by extreme divergence of the knees. The nymphæ cannot be made to part by any posture, and have to be artificially opened in order to expose their inner surface. The vulvar slit is vertical and mesial, while that of the vagina is transverse. The ostium vaginæ can be seen only after the labia majora and minora have been separated. The line between skin and mucous membrane is quite sharply defined in the genital organs of the living female. It extends continuously along the base of the inner surface of each labium minus, the base of the outer aspect of the hymen, and beneath the prepuce of the clitoris.

The labia majora are firm and full in young adults with perfect health; but, in the old or debilitated, they tend to become wrinkled and pendulous from a decrease in their adipose tissue. The term "vulva" was applied by the ancients to the labia from a fancied resemblance to a valve to the vaginal orifice (valva). When the labia are full and rounded, the term "*vulva connivens*" is applied; when they tend to gape from their flaccidity (the subject being in the supine posture), the term "*vulva hians*" is used to designate their condition. The two extremities of the labia are designated by most anatomists as the "*anterior*" and "*posterior commissures of the vulva*." Luschka has shown that these terms are incorrectly applied since the labia are continous, in front, with the mons Veneris, and, behind, with the perineal structures.

The *labia minora* are two reddish folds of mucous membrane, which lie between the labia majora, and with whose internal surface they are attached at their outer borders. They are naturally moist, since their covering partakes of some of the characters of a mucous surface, in which respect they differ from the labia majora, although it is classed as skin by most observers. They form a covering for the clitoris, which is analogous to the prepuce of the male, by splitting into two folds which surround that organ, and help to complete the suspensory ligament of the clitoris. In the young virgin, these folds are completely concealed by the labia majora; but, when the vulvar opening gapes, and becomes pendulous, the contact of air causes the labia minora to assume a dirty, bluish color, and to partake of the character of integument. In the Hottentot and Bushman women, these folds often become of excessive length; they may even reach the condition called the "Hottentot apron," where they hang upon the thighs. The nymphæ are supposed to assist in directing the flow of urine, and in bringing the clitoris in close approximation with the penis, as they are separated and forced inward during coition. The "fourchette" is considered by some anatomists as a direct continuation of these folds.

Scattered hairs are found upon the surface of the labia minora of the adult.

THE URETERS.—These tubes convey the urine from the renal pelvis, which may be regarded as the expanded portion of the ureter, to the bladder. Their topographical relations are of great importance; because an accurate knowledge of their course in the non-pregnant and pregnant female is necessary for the proper performance of certain surgical operations— chiefly those for the repair of vesico-vaginal fistulæ, and of gastro-elytrotomy and extirpation of the uterus. Among those who have devoted special attention to the subject may be mentioned Freund, Garrigues, Pawlick, and Polk. The points which seem to be those of controversy between the authors mentioned are (1) the exact relations of the point of entrance of the ureter into the cavity of the bladder to the uterine cervix and the internal orifice of the urethra; (2) the variations in the general course of the ureters produced by the enlargement of the gravid uterus; (3) the bony points which may be used as guides to these tubes during their passage through the pelvis; (4) the guides to catheterism of the ureters; and (5) the relations of these tubes to the vagina and broad ligaments.

The original investigations of Garrigues [1] were made upon the non-pregnant female, and are in accord, in most essential particulars, with the researches of Freund, who studied this subject with special reference to the perfection of the operation devised by him for extirpation of the uterus. Both of these observers have corrected and supplemented the statements made by Sappey, Luschka, and Savage, respecting the normal course of the ureters. As I am impressed from my own observations with the accuracy and completeness of some of Garrigues' deductions, I take the liberty of quoting from his paper, as follows:

"The ureter is the continuation of the renal pelvis. It lies behind or under the peritonæum, imbedded in very loose connective tissue, and is much longer than the direct line between its two ends. The left ureter begins somewhat higher up than the right. The distance from the starting-point of the right

[1] Gastro-Elytrotomy, N. Y. Med. Jour., Oct., and Nov., 1878. The same author has published subsequent measurements (made on the cadaver of a parturient woman who died a few days after the expiration of the full term of pregnancy) which partially confirmed those of Polk, AM. JOUR. OBSTET., Jan., 1883. It must be evident to all, however, that the cadaver of a woman who died *after delivery* is not as reliable a standard for measurements as a pregnant cadaver, if the determination of the position of the ureters during gestation is the question at issue.

ureter from the renal pelvis horizontally to the left ureter I have found to be two inches and a half (6.4 centimetres). From this point they go, excepting slight windings, paralled with one another down to the spot where they cross the iliac vessels, so that the distance between them at this latter point is likewise two and a half inches (6.4 centimetres). In this part of their passage they lie in front of the psoas muscle. The left ureter crosses the iliac vessels somewhat higher up than the right, the left lying in front of the lowest part of the common iliac artery, the right in front of the uppermost part of the external iliac artery. The right passes also in front of the external iliac vein, which here lies outside the artery. Consequently the left is also slightly nearer to the median line of the vertebral column. From this point they diverge, running downward, backward and a little outward, on the wall of the pelvis to a point near the *spina ischii*, at which point they are farthest separated from one another, namely, about three inches and a half (8.5 centimetres). The ureter lies outside the hypogastric artery. They run behind the broad ligaments down to the indicated point near the spina ischii, and bend then downward, forward, and considerably inward, so as to converge toward the bladder. They pass beneath the base of the broad ligaments, lying in the abundant cellular tissue found in this locality. They cross the cervix at some distance from behind, at an acute angle, so as to come in front of it and below it. They lie outside and above the anterior part of the side-wall of the vagina, if we will suppose such a thing to exist, on a spot as large as the tip of the finger. On reaching the wall of the bladder, they turn rather sharply inward and go less downward, until they open with a small slit in the interior of the bladder, at the outer angle of the *trigonum vesicale*. From behind they are seen to be united by a kind of ridge forming the base of the trigonum."

The same author then gives some measurements which are of value to the operative surgeon. These are embraced in the following quotation:

" I found the following distances: From the ureter horizontally to the horn of the uterus (the starting-point of the ovarian ligament), right, three-quarters of an inch (1.9 centimetre), left, an inch and one-eighth (2.9 centimetres); from the ureter, horizontally, to the junction of the body and the neck of the womb, right, three-quarters of an inch (1.9 centimetre), left, five-eighths of an inch (1.5 centimetre); from the ureter to the nearest point of the vaginal portion on either side, half an inch (1.3 centimetre); from the ureter where it enters the wall of the bladder to the utero-vaginal junction, right, three-quarters of an inch (1.9 centimetre), left, five-eighths of an inch (1.5 centimetre); this point is on a level with the os uteri; from the opening into the cavity of the bladder to the vaginal portion, right, half an inch (1.3 centimetre), left, five-eighths of an inch (1.5 centimetre); between both ureters, where they reach the wall of the bladder, two inches (5 centimetres); between the two openings of the ureters into the cavity, one inch (2.5 centimetres); from the point where the ureter passes under the broad ligament to the point where it reaches the wall of the bladder, one inch and one-eighth (2.9 centimetres); the course of the ureter in the wall of the bladder, nine-sixteenths of an inch (1.4 centimetre); from the opening of the ureter into the cavity of the bladder to the centre of the anterior lip of the os, right, three-quarters of an inch (1.9 centimetre), left, one inch (2.5 centimetres); these openings lie below the level and considerably in front of the vaginal portion; from the internal opening of the urethra to the os uteri, one inch and a quarter (3.2 centimetres)."

The deduction of Garrigues, that the vaginal incision in gastro-elytrotomy should be performed *below the ureter*, has been made a point of issue between that observer and Polk. The latter has confined his investigations to pregnant women, and has, in consequence, been led to some conclusions which differ from those of the author quoted. He believes that the ureter should lie below the point of extraction of the fetus in gastro-elytrotomy rather than above it, and hence below the seat of the vaginal incision, since experiments upon the cadaver showed that the ureter suffered laceration in one instance from extreme tension, when the vaginal incision was made below it.

In a later article, Garrigues[1] has acknowledged, however, that the incision, as suggested by Polk, may be made above the ureter with greater safety to the patient.

[1] AMERICAN JOURNAL OF OBSTETRICS, January, 1882.

The following quotation from the article of Polk[1] will show the points of variation from the non-pregnant standard which the ureters undergo in consequence of the altered position of the uterus during the latter months of gestation. He says:

"In the first place, I have found that the ureters do not follow the pelvic wall to a point near the ischial spine, as in the non-pregnant condition. I have found their courses to be as follows: Crossing the pelvic brim at the common iliac bifurcation, the left just behind, the right just in front, of that point, they descend into the canal to the brim of the bony pelvis, the point being about the synchondrosis. In this course, they accompany the internal iliac artery, the right in front of the vessel, the left crossing it obliquely. Reaching the bony brim (the ilio-pectineal line), they leave the pelvic wall, emerging from beneath the base of the broad ligaments (in pregnancy about on a level with the pelvic brim, and carried back on a line with the synchondrosis), and take a course downward, forward, and somewhat inward, passing about midway between the pelvic wall and the cervico-vaginal junction, but approaching very closely the antero-lateral wall of the vagina, as they turn more decidedly inward, on a lower plane, to strike the base of the bladder three-quarters of an inch below the cervix, terminating finally in the bladder at a point (the subject being on the back) just two inches below the spine of the pubes.

"A line drawn from the bifurcation of the common iliac to the spine of the pubes, I have found to correspond in the main to the line of the ureters. Along this line they have the following relations to the pelvic brim (in the recent state): At the bifurcation, half an inch below;[2] at the extremities of the transverse diameter of the pelvis, about an inch; and at the spine of the pubes, two inches below. As a whole, the tubes in the pelvis are situated upon a higher plane than in the non-pregnant condition, having been carried slightly upward while being separated from their close relations with the pelvic wall by the

---

[1] New York Medical Journal, May, 1881.

[2] This would *seem* to be an error (as first pointed out by Garrigues in his article of January, 1883), but Polk has explained that his measurements were made from the anterior edge of the psoas muscle, and that the term "below" used by him corresponds to the antero-posterior measurement in the standing posture of the woman.

ascending uterus. How far they may be elevated in a case of extreme pelvic deformity with a pendulous abdomen, with the uterus correspondingly displaced, I am unable to say; but I think it probable that, the bladder being empty and not dragged upward, thus preserving the normal position of the vesical end of the tubes, the displacement would not be such as to bring any part of them much above the points indicated."

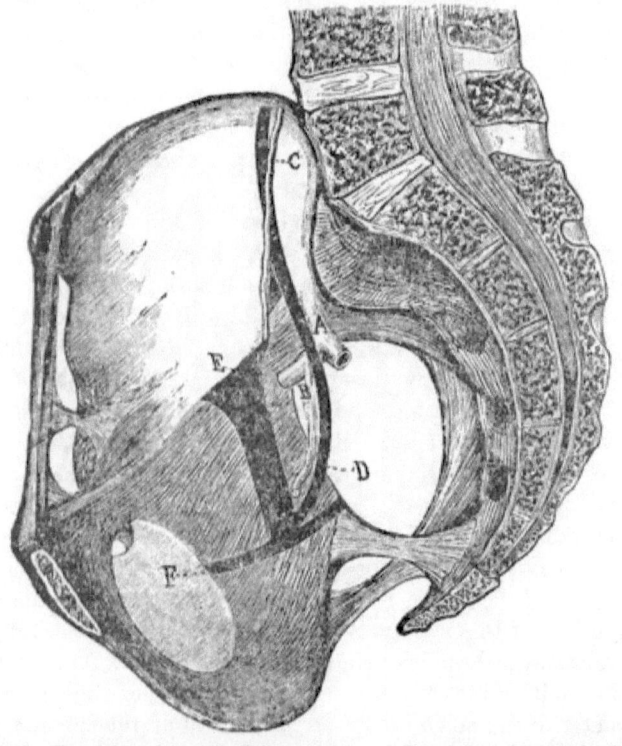

FIG. 15.—The right pelvic wall (Polk). A, internal iliac artery, its anterior trunk turned aside to show origin of uterine artery; B, uterine artery; C, ovarian artery; D, course of the ureter, projected on pelvic wall; E, line of pelvic attachment of the broad ligament of uterus in the nulliparous female; F, line of attachment of the levator ani muscle, marking the level of the base of the broad ligament.

These quotations will enable the reader to more readily contrast the latest researches of these investigators, and to form a definite conception of the course of the ureters in any case demanding a surgical procedure in which they are liable to be injured. It will be perceived that Polk designates certain bony and other fixed points which may prove valuable surgical guides to the ureters in different portions of their course. He also establishes the fact that the tubes lie on a *higher plane in the pregnant state* than in the non-pregnant condition.

In the parturient cadaver examined by Garrigues (subsequent to his published measurements previously quoted), it was found that the point where the ureters crossed the common iliac arteries was situated one and five-sixteenth inches above the brim of the pelvis; at the end of the transverse diameter of the pelvis, the ureter lay two inches below in a vertical line (the subject lying upon her back); and, from the spine of the pubes, a

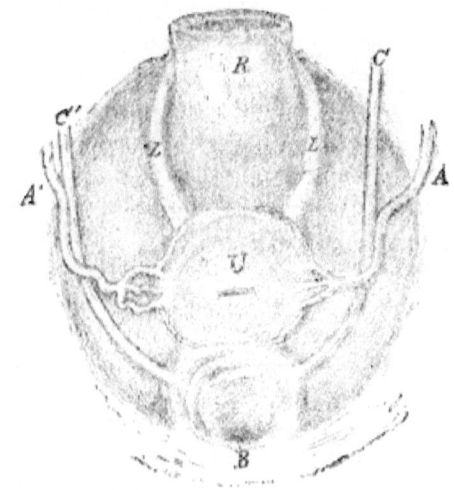

Fig. 16.—The floor of the pelvis (Polk). The uterus and broad ligaments, as well as the bladder, have been cut away on a level with the internal uterine os. Reproduced from a photograph. U, uterus; B, bladder; R, rectum; A, A', uterine arteries; C, C', ureters; L. utero-sacral ligaments.

straight line to the point of the junction of the ureter with the bladder measured three inches. These measurements differ from those of Polk, quoted in a previous page. Polk found, moreover, that the ureter reached the base of the bladder at the level of the os uteri. Garrigues found it to be three-quarters of an inch below the cervix; but he attributes this variation to an empty bladder, since Luschka places it at the junction of the upper and middle thirds of the anterior wall of the vagina when the bladder is distended (as shown in his drawing).

Subsequent investigations upon pregnant cadaver (unfortunately very rare when pregnancy is far advanced) will help to positively determine many anatomical points pertaining to the operation of gastro-elytrotomy, which must still be considered as unsettled. As yet, there have been no published measurements made (to my knowledge) upon the *pregnant* cadaver to confute those of Polk.

The alterations in position and attachment which the broad ligaments of the uterus undergo, in consequence of the growth of that organ during gestation, shed much light, through the investigations of Polk, upon the relations which the ureters bear to these ligaments. In the nulliparous subject, the base of the broad ligament of either side rests upon the fatty tissue between the vagina and the recto-vesical fascia, which covers the upper surface of the levator ani muscle; hence the pelvic peritoneum sinks deeply in the pelvis at the sides of the uterus, and hugs the pelvic wall along a vertical line situated between the sciatic notch posteriorly and the obturator foramen anteriorly. During gestation the uterus drags the broad ligaments upward in proportion to its increasing size; so that at full term the bases of these ligaments lie on a level with the pectineal line, and extend from the pectineal eminence, anteriorly, to the synchondrosis, posteriorly (these limits being determined by the round ligaments of the uterus and the ovarian arteries). The separation of the laminæ of the broad ligaments gives to that of each side a *triangular form* at the full term of pregnancy, the base of the triangle corresponding to the pectineal line, and its apex to the horn of the uterus. When parturition has been accomplished, the broad ligaments tend to slowly regain the position within the pelvic cavity which they originally occupied. Now, it can be readily understood why the ureters can be said to have no fixed relations with the broad ligaments, because they do not suffer the same displacement during the pregnant state. In the virgin and nullipara, they hug the pelvic wall in a plane posterior to the pelvic attachments of these ligaments. In the subject far advanced in pregnancy, they have little if any relation with them. In subjects who have borne children, they may lie between the laminæ of these ligaments. Garrigues has taken issue with Savage because the latter author has described the ureters as passing between the layers of the broad ligaments. Polk has shown that they may and may not follow such a course; thus sustaining the accuracy of both observers.

Catheterism of the ureters is sometimes demanded, chiefly as

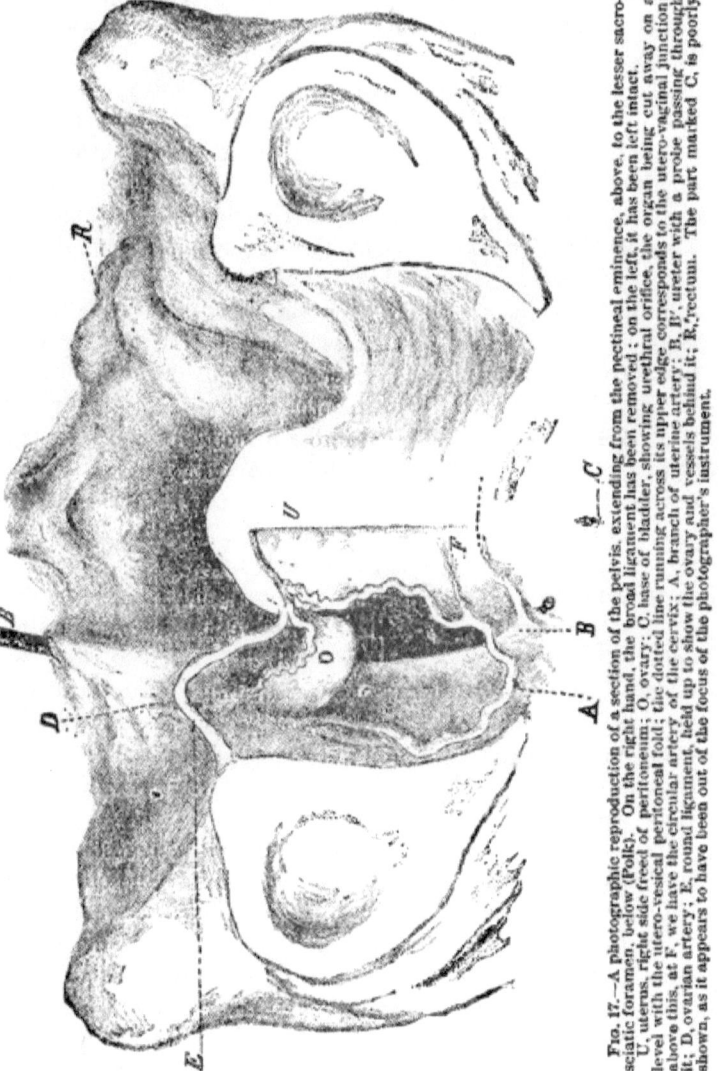

Fig. 17.—A photographic reproduction of a section of the pelvis, extending from the pectineal eminence, above, to the lesser sacro-sciatic foramen, below (Polk). On the right hand, the broad ligament has been removed; on the left, it has been left intact. U, uterus, right side freed of peritoneum; O, ovary; C, base of bladder, showing urethral orifice, the organ being cut away on a level with the utero-vesical peritoneal fold; the dotted line running across its upper edge corresponds to the utero-vaginal junction; above this, at F, we have the circular artery of the cervix; A, branch of uterine artery; B, B', ureter with a probe passing through it; D, ovarian artery; E, round ligament, held up to show the ovary and vessels behind it; R, rectum. The part marked C, is poorly shown, as it appears to have been out of the focus of the photographer's instrument.

a step toward determining their outline and course during surgical procedures. It has been customary to dilate the urethra for this purpose. Pawlick, of Vienna, however, has lately described a new method of exploring the ureters by which he claims that all preliminary operations may be dispensed with. He states that when a woman is placed in the genu-pectoral posture and the perineum is retracted, two furrows may be discovered on the anterior wall of the vagina, which, starting from a common point situated slightly behind the bulge of the urethra, diverge at an obtuse angle, and pass backward and outward. In the region of the cervix uteri a transverse furrow may also be seen to connect the extremities of these diverging furrows, thus forming a triangle which corresponds with the situation of the trigonum. He believes that the *diverging furrows* correspond with the line of the ureters, and that the orifice of each ureter may be found at or near to the point of intersection of the transverse and diverging furrow of the corresponding side. After a sound is introduced within the bladder, this author states that it is an easy matter [1] to guide the point of the sound into the mouth of the ureter with the finger in the vagina, if this guide be employed. This procedure is of especial importance in the operation for extirpation of the uterus, and the suggestion of Pawlick is worthy of a thorough trial. I have not been able to personally test it to my satisfaction, as the furrows have not generally been distinct, even in those cadavers to which I have had access soon after death. In some cases I could not detect them at all. My friend, Professor Polk, has assured me that he has been unable to verify the value of Pawlick's suggestion in his attempts made upon the living subject; and that the cadaver usually fails to bring the guides into prominence, although they can be easily recognized during life.

THE PELVIC PERITONEUM.—If, as Henle suggests, we suppose that those pelvic organs which rise above the plane of the superior strait were thrust upward during their development to reach their proper position, and thus were made to displace the peritoneum which naturally sought a lower level, we can form a clearer insight into its reflections than without some such homely illustration. In several pages which have preceded, the

[1] I doubt if any one who has attempted the step will agree with the author regarding its *ease* of execution, even if it has been successfully performed. The suggestion is as yet too new to be sustained as of great value or to be rejected as worthless.

reflections of the pelvic peritoneum have been discussed[1] — chiefly in connection with the uterus and bladder—and some of the cuts and diagrams incorporated will prove of the greatest assistance now in reviewing this important portion of topographic pelvic anatomy. The reader is referred, therefore, to all representations of sagittal sections of the pelvis previously given, and to the diagrams of Hodge on pages 46 and 47, should any portion of the description seem vague or incom-

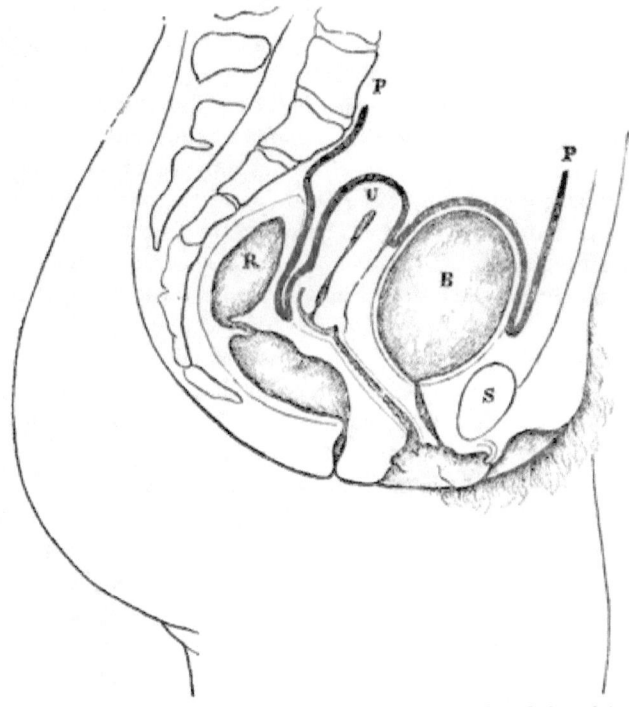

Fig. 18.—A diagram designed to show the antero-posterior outline of the pelvic peritoneum in the mesial pelvic plane.

P-P, peritoneum; R, rectum; U, uterus; B, bladder when distended; S, symphysis pubis; the *vesico-abdominal* and *urethro-vesical pouches* and *Douglas' cul-de-sac* are made very apparent.

plete. Hart and Barbour have lately published a work in which modifications in the appearance of the peritoneal reflections are made very clear by reproductions of several sagittal frozen sections, selected from other authors.

Savage very wisely remarks that the greater portion of the

[1] Pages 44, 45, 46, 47, 48, 51, 53, 61.

surgical interest which pertains to the peritoneum depends for the most part upon its substratum. The membrane itself consists simply of a delicate and highly elastic basement structure, with a covering of epithelium; and it certainly seems to be incapable of exerting any marked ligamentous action upon the organs with which it comes in contact. It is bound to those structures with which it bears relation by a layer of connective tissue—never entirely wanting—which is developed in inverse ratio to the firmness of its union to the parts so covered. The substratum to which Savage refers comprises certain involuntary muscular fibres which form a layer, which that author designates as a "platysma muscle," over the uterus and its appendages and the ligaments connected with them. This muscular lining of the pelvic peritoneum may be traced along the Fallopian tubes and between the folds of the broad ligaments. The sacro-uterine and round ligaments are abundantly supplied with muscular fibres; some of which are derived, in the case of the former, however, from the cervix uteri and also from the vagina.

The round ligaments of the uterus, which seem to be a continuation of the muscular structure of that organ, are supplied, in part, from this platysma muscle. The utero-ovarian ligaments have a similar supply of muscular fibres which are prolonged from a set which accompany the spermatic vessels.

The elasticity of the peritoneum enables that membrane to undergo changes, as to its relations and dimensions, which are truly remarkable. The increase in size of the bladder during its full distention, and of the uterus during gestation, seem to produce no corresponding alteration in its fixed attachments. Savage tried the experiment of distending the bladder through the ureters (the urethra having been previously tied) until it rose above the level of the umbilicus; yet he found that the vesico-pubic fold of the peritoneum still extended to within one inch of the upper border of the symphysis.[1] We have reason to think that the pouch of Douglas, which extends between the rectum and the vagina, is seldom raised to any

---

[1] Hart and Barbour do not appear to fully coincide with this deduction. There seems to be no doubt that, in the *genu-pectoral* posture, the peritoneum passes from the anterior abdominal wall to the fundus of the empty bladder about one inch and a half above the symphysis pubis.

appreciable extent during gestation. It has been demonstrated, however, by Polk, that the *base of the broad ligaments are raised* to nearly the level of the brim of the pelvis at the full term of pregnancy, and their direction so altered as to cause their attachment to the bony pelvis to approach the neighborhood of the synchondrosis. In the operation of gastro-elytrotomy, it has, moreover, been demonstrated positively that the peritoneum can be easily stripped from its normal attachments in the neighborhood of the brim of the pelvis, and raised to such an extent as to admit of the extraction of a full-sized fetus beneath it.

Since the peritoneum, like all serous membranes, is a closed sac, which communicates directly with nothing except the lymphatic vessels, that probably open into its cavity by innumerable pores, it follows that foreign products within its cavity, such as blood, pus, inflammatory exudation, etc., can be removed only by the process of absorption or by actual disintegration of its elements.

In the diagrammatic cuts of Hodge which have been incorporated in previous pages of this article, it will be seen that the peritoneum is spread over the pelvic viscera from the line of its reflection from the wall of the abdomen and the inner surface of the false pelvis. When the bladder is empty, the uterus projects higher above the plane of the symphysis than any other pelvic organ, save the rectum. It appears, therefore, in most of the antero-posterior median sections of the pelvis to have an excess of peritoneal covering when compared with the bladder, in front, and the rectum, behind; since the whole of the posterior surface, the fundus, and the upper three fourths of the anterior surface is invested by it. From the sides of the uterus two folds of the peritoneum are reflected to the lateral wall of the pelvis—forming the broad ligaments of that organ. These broad ligaments are usually described as presenting three minor folds—the anterior, middle and posterior—wich embrace respectively the round ligament, the Fallopian tube, and the ovary of each side. The external margin of the broad ligament is normally attached to the lateral wall of the pelvis along a line which is situated between the great sacro-sciatic notch and the margin of the obturator foramen as far down as the level of the ischial spine; its base rests upon the connective tissue separating the vagina from the recto-vesical fascia which covers the levator ani muscle; its internal margin is attached to the side of the uterus; its upper or free margin is somewhat concave and extends from the side

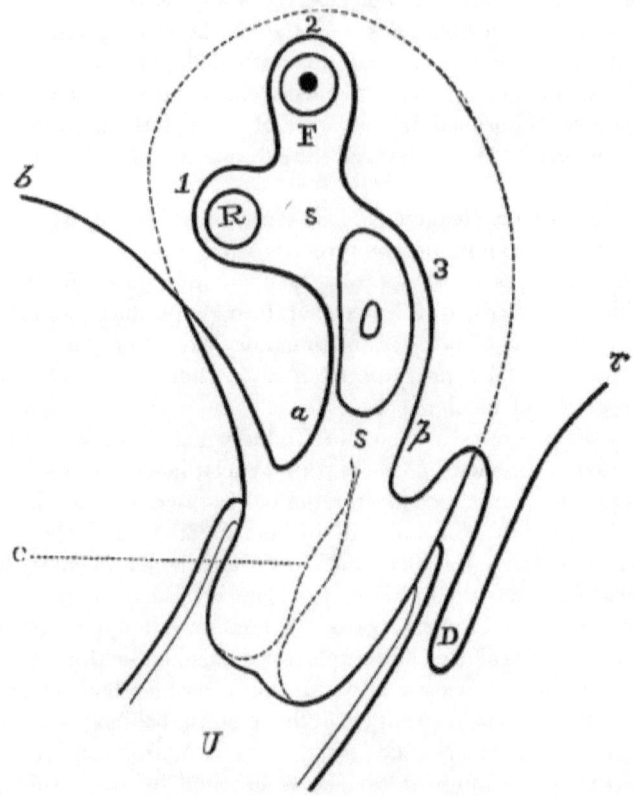

Fig. 19.—A diagram designed to show the three minor folds of each broad ligament of the uterus, and the structures which are contained between its two layers; as seen, in a lateral sagittal pelvic section, cutting the ligament close to the border of the uterus.

1, 2, 3, anterior, middle, and posterior folds; R, round ligament; F, Fallopian tube; O, ovary, its anterior border lying in close contact with the peritoneum; V, vagina; D, the pouch of Douglas; c, outline of cavity of the cervix; a, anterior layer of broad ligament; p, posterior layer of same; b, reflection of peritoneum to bladder; r, reflection of same to rectum; s, space filled up with blood-vessels, nerves, connective tissue, lymphatics, and muscular tissue.

of the uterus to the lateral wall of the pelvis at a point on the pectineal line, situated in the virgin about midway between the sacro-iliac synchondrosis and the ilio-pectineal eminence.

[1] After the female has once reached the full term of pregnancy, this point may approach nearer to the synchondrosis; as the broad ligaments

These ligaments, upon either side of the uterus, in connection with the uterus itself, form a septum which divides the pelvic cavity into an anterior and posterior space. In the former, the outline of the bladder is perceived, from a superior view of the pelvis, beneath the peritoneum, provided that the organ is distended; while in the latter, the rectum is seen escaping between the sacro-uterine ligaments which pass to either side of that tube to reach their sacral attachments.

The free margin of the broad ligament is made prominent by the presence of the Fallopian tube beneath it. With the growth of the uterus during gestation, the base of this ligament is gradually raised, until it reaches nearly to the level of the pectineal line at full term.[1] At the same time, the attachment of the superior border is gradually carried forward to the pectineal eminence and backward toward the synchondrosis, thus giving them a distinctly triangular form, the apex of which corresponds to the horn of the pregnant uterus. After parturition, the broad ligaments appear to slowly regain their normal relations to the pelvic structures.

Between the uterus and the vagina in front, and the rectum behind, may be seen a pouch of peritoneum which dips downward into the pelvic cavity far below the level of the other reflections of that membrane.[2] This is commonly called "*the pouch of Douglas.*" It is an important surgical space, since fluid within the peritoneal cavity would naturally gravitate there and be discoverable by vaginal examination; while intestine, a retroverted or retroflexed womb, an impregnated ovum, a displaced ovary, and new growths of the uterus itself might possibly be detected in this region. The extent to which this pouch commonly descends into the pelvic cavity has been given in a foot note on a previous page. Like all serous pouches, it consists of two folds of the peritoneum.

From the anterior surface of the uterus, the peritoneum is prolonged to the bladder, thus forming the so-called *utero-ves-*

are carried backward by the uterus as it enlarges, and slowly again their normal position.

[1] The late researches of Prof. W. M. Polk have added much to our knowledge of this subject.

[2] Pirogoff shows, in his frozen section, a descent of this pouch almost to the ostium vaginæ. This is, of course, an extreme limit, but is a suggestive abnormality to the operating surgeon.

ical ligaments, and, from the bladder, it passes to the anterior abdominal wall. When the bladder is distended, the reflection of the peritoneum from that viscus to the abdominal wall forms a pouch—the *pubo-vesical pouch*—which reaches to within one inch of the pubes, or even less.

The *round ligaments* of the uterus appear, through the investing peritoneum, as two curved ridges which pass from the sides of the uterus, in front of and below the attachments of the Fallopian tubes, to the internal opening of the inguinal canal. They lie therefore to the outer side of the bladder

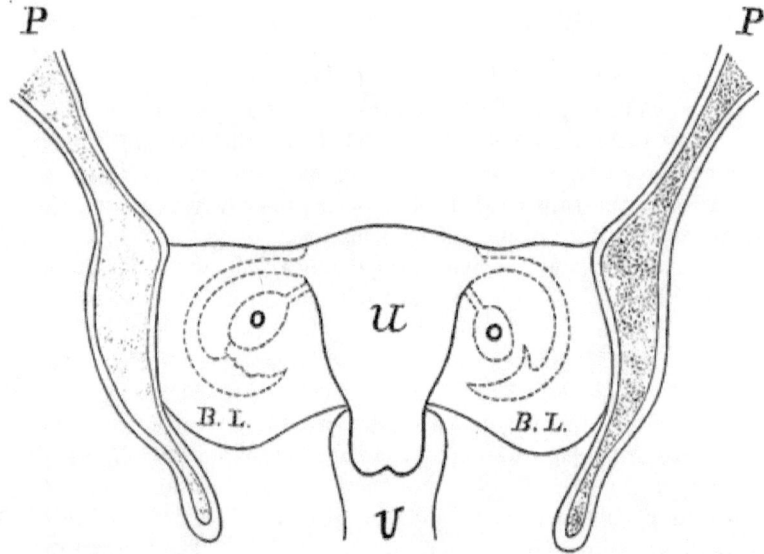

FIG. 20.—A diagram designed to show the attachments, outline, and some of the relations of the broad ligaments of the uterus.
P, pelvic bones; U, uterine body; V, vagina; O, ovary; F, Fallopian tube; BL, broad ligament. Note that the ovaries are represented as more nearly vertical than in most cuts, and that the Fallopian tubes are curved.

when distended with urine, but near to the pelvic wall. Within the inguinal canal, a prolongation of peritoneum accompanies them—the *canal of Nuck*. This sometimes remains open and gives rise to hydrocele of the inguinal canal.

By reference to Fig. 10 it will be seen that the utero-sacral ligaments are given off at a lower level than the broad ligaments; in fact they correspond nearly to the level of the anterior wall of the vagina (as may be seen in Fig. 6) of which they are a practical extension and with which they are physiologically associated.

The situation of the ovary within the broad ligament of the uterus is on a lower plane than that of the Fallopian tube or the round ligament. The Fallopian tube encircles it, as it were, and the fimbriated extremity of that tube curves to reach the ovary. The disputed points pertaining to the relations between the ovary and the Fallopian tube will be discussed later.

The uterus is the most movable pelvic organ of those in relation with the peritoneum. The broad ligaments unquestionably assist to some extent in preventing its lateral and anteroposterior displacements.[1] The round ligaments tend possibly to prevent its backward displacement, but they are too lax to have any marked influence upon the womb. The utero-sacral ligaments act as an aid to the anterior vaginal wall in supporting that organ. Luschka has advanced the view that the muscular fibres of the utero-sacral ligaments assist in drawing the cervix backward and thus tend to cause anteversion of the fundus. I am inclined to differ from this view. The muscular fibres of these ligaments are so attached to the uterus as to be a practical extension of the fibres of the anterior wall of the vagina; hence if they contract they are opposed by the anterior vaginal wall, unless the uterus be drawn downward below its proper level, when both will act together and tend to replace it. The anterior vaginal wall in front, the muscular structure of the uterus itself which intervenes between the vagina and the utero-sacral ligaments, and the sacro-uterine ligaments behind, form what may be regarded as a *supporting beam* (Foster) which runs from the pubes to the sacrum, and the fibres of each are so closely intermingled as to be properly regarded as continuous. If the musclar fibres of the utero-sacral ligaments act alone, the anterior vaginal wall is so placed as to resist their action (unless it be abnormally relaxed); since it is made

---

[1] These ligaments (being attached to the entire length of the body of the uterus) constitute in the nulliparous female an elastic septum which lies in the plane of the transverse pelvic diameter. Although lax, they thus tend to some extent to keep the organ erect, and afford an explanation of the lever-like action of vaginal traction upon the attitude of the womb (see page 27 which treats of the Hodge pessary).

tense in health in order to support the bladder which rests upon it.

Special points of interest pertaining to the pelvic peritoneum.—Some of the statements of previous pages may be thus summarized:

The *vesico-abdominal pouch* exists only when the bladder is distended. In the collapsed state of that organ the peritoneum passes directly over the fundus below the level of the horizontal plane which intersects the upper border of the symphysis. The depth of this pouch is probably modified also by the increased size of the uterus during gestation, irrespective of the distention of the bladder, as the peritoneum is lifted. This pouch contains no small intestine, either when the bladder is in systole or diastole (Hart). The view upheld by Savage that the point where the peritoneum leaves the anterior wall of the abdomen is not subject to change, is apparently confuted by some frozen sections which have appeared.

A *utero-abdominal pouch* exists when the bladder presents the Y-shaped section, as the peritoneum is then reflected directly from the anterior wall of the abdomen upon the anterior surface of the uterus, without any vesical elevation. In this case, small intestine fills the intervening space.

The *pouch of Douglas* has the following boundaries: Above it is limited by the utero-sacral ligaments of either side; below, it terminates commonly at the vaginal fornix, although Pirogoff has shown in his section an extreme depth of this pouch; anteriorly, it lies in relation with the fornix vaginæ, and the supra-vaginal portion of the cervix; posteriorly, it lies in relation with the rectum. Hart claims that it normally contains intestine, but this view is not generally accepted. Its intimate relations with the cervix, the posterior wall of the vagina, and the rectum, gives it a surgical importance in excess of other peritoneal reflections. Being the most dependent part of the peritoneum, it has also a special importance in the diagnosis of pelvic effusions, displacements of the ovary, abdominal pregnancy, etc.

The peritoneum forms the *broad ligaments of the uterus*, which bind that organ to the lateral walls of the true pelvis. Between the folds of this peritoneal reflection, the Fallopian tubes of either side are found at its upper free margin, as well as the ovarian artery : the ovaries lie in the posterior folds, but are attached to the anterior laminæ; the round ligaments of either side run in the anterior folds in a curved direction ; in the region of its base, the uterine artery of either side is found; finally, the ureters of either side pass beneath its base, but do not, as a rule, lie between its laminæ.[1] The attachments of this peritoneal fold to the side-wall of the pelvis in the nulliparous female has been admirably shown by Polk in a cut devised by him (Fig. 15). My own diagram (Fig. 19) will also help to interpret the arrangement of its laminæ.

These ligaments are not drawn tight, so as to act as lines of tension upon the uterus, but are quite lax in the non-pregnant female. *Gestation*, however, causes *marked alterations* in their position and bony attachments (Polk).

The cellular tissue found between the laminæ of the broad ligaments has been described by Guerin as distinct and separate from that of other regions of the pelvis ; a statement of importance if true, but one not generally accepted.

Displacements of the uterus cause coincident displacement of the broad ligaments. Cicatrices of these ligaments cause unilateral deviations of the uterus.

The blood-vessels, lymphatics, muscular fibres, and connective tissue found between the laminæ of the broad ligaments suggest some practical deductions, as follows : The two large arteries found between its layers (ovarian and uterine) lie at the extreme upper and lower limits of the ligament and are derived from different sources. Both supply the uterus with blood and anastomose freely ; hence, abdominal pressure upon the aorta in uterine hemorrhage would tend to only partially arrest the flow of blood, provided the spermatics were not controlled. Fortunately, the calibre of the uterine artery is greatly in excess of the ovarian in the pregnant state. The uterine artery bears a relation to the vaginal wall before it reaches the uterus ; a fact to be remembered in making incisions in that tube at its upper part, especially in the pregnant woman.

The *pampiniform plexus* of veins derives its blood from the uterus, ovaries, and Fallopian tubes, and its vessels coalesce into

---

[1] Polk has shown that only when these ligaments are attached near to the synchondrosis does the ureter pass between the two laminæ. This is stated by Savage to be the normal course of these tubes.

a single trunk, the internal spermatic vein (utero-ovarian vein—Sappey) which follows the course of the artery of the same name. This plexus is one of the most frequent seats of rupture in pelvic hematocele. It is important to remember that the veins of the uterus, vagina, bladder, urethra, and rectum anastomose freely with each other and with the veins of the perineum—constituting what may practically be considered as a continuous chain of vessels. This explains why, in pregnancy, when the veins of the pelvis are engorged, small wounds of the generative organs are sometimes associated with dangerous and even fatal hemorrhage.

The *lymphatic vessels* of the broad ligaments are derived from the borders and fundus of the uterus; after following the general course of the pampiniform plexus of veins, they unite with the lymphatics of the lumbar region. Leopold has pointed out that the lymphatics of the cervix follow a different course from those of the fundus and borders of the organ, since they empty into the glands of the pelvic cavity.

The *connective tissue* between the layers of the broad ligaments serves, as in other parts of the pelvis, to unite the organs with which it is connected to the pelvic walls, and also to fix them in their proper relations with adjoining structures. The physiological and clinical suggestions afforded by this type of structure will be considered later.

The modifications in the shape and attachments of the broad ligaments which are produced during gestation, have been referred to under the discussion of the ureters.

Most of the reflections of the peritoneum, thus far summarized, have been suggested by frozen sections of the pelvis, made through its different planes. Let us now consider the points of interest suggested by a superior view of the female pelvis in the recent state.

The uterus and its broad ligaments divide the pelvis into an anterior and posterior fossa (see page 46). In the former, the peritoneum is loosely attached throughout its entire extent. Where it is reflected from the anterior face of the uterus to the bladder, it hangs so loosely that, by a very slight elevation of the uterus, it may be brought into contact with the upper wall of the vagina. This fact should be borne in mind when surgical procedures are attempted upon the anterior lip of the cervix, when raised above its normal plane.

The posterior fossa of the pelvis may be considered as consisting of two compartments: An upper, lying above the level of the sacro-uterine ligaments, called by Polk the "retro-ovarian shelves;" and a lower, the pouch of Douglas. The floor of the upper is broken by an opening, between the sacro-uterine ligaments, which enters into the lower. The form of the upper has been compared to two right-angled triangles united at their bases—the sacro-uterine ligaments; the other sides being formed by the base of the broad ligament of either side, and the pelvic wall. Upon the floor of this compartment or shelf, the ovaries rest when slightly prolapsed. The dimensions of this shelf are usually less upon the left side than the right, on account of the presence of the rectum; occasionally, however, they are symmetrical. The plane of the floor intersects the pelvic wall just below the great sacro-sciatic foramen.

Throughout the entire posterior fossa of the pelvis, the peritoneum is more closely attached to adjacent structures than in the anterior fossa.

The floors of both the anterior and posterior fossæ of the pelvis (save in the case of Douglas' pouch) seldom, if ever, fall below a plane which intersects the central point of the symphysis and the junction of the third and fourth bones of the sacrum. During advancing pregnancy, however, the lowest level is raised gradually till it reaches a plane which intersects the centre of the symphysis and a point just below the promontory of the sacrum (Polk). The highest limit is reached about the thirty-sixth or thirty-eighth week of gestation. At this time, the posterior fossa of the pelvis (except the pouch of Douglas), is entirely obliterated, as the floor of the "retro-ovarian shelves" lies above the brim of the pelvis; while the anterior fossa of the pelvis is rendered shallow. Polk states that the backward displacement of the broad ligaments diminishes also the antero-posterior and lateral measurements of the posterior fossa.

The pouch of Douglas is somewhat affected by the upward stretching to which the vagina is subjected during gestation, but the alteration in the level of its floor is small compared to that of the "retro-ovarian shelves." The sacro-uterine ligaments change their attachments to the sacrum[1] as the uterus enlarges—rising as high as the first bone, near to the promontory—and, as they mark the upper limits of the cul-de-sac, we find it converted into a deep and rather narrow channel, about large enough to contain a moderately filled rectum (Polk).

The pelvic peritoneum is in intimate relation with most of the large blood-vessels, which are distributed to the uterus, ovary, vagina, and bladder, and also with plexuses of smaller vessels which ramify in the subserous tissue of the pelvis. We are forced to admit two varieties of hematoma of the pelvic region, viz., subperitoneal or "cellular hematoma," and intraperitoneal extravasation, which are properly termed "hematocele." Savage mentions a long list of cases which resulted in death from one of these two causes. It seems to be believed by some authors, on purely anatomical grounds, that when a tumor, due to extravasations of blood, reaches an *extreme size*, it must be considered as intraperitoneal; since the fixed relations[2] of the peritoneum would apparently preclude any enormous escape from the blood-vessels beneath it, without the peritoneum were itself lacerated (Savage). In deciding as to the character of any such tumor, a layer of plastic exudation, if present upon the surface of an incapsulated intra-pelvic hematoma, might tend to mislead the examiner as regards the true situation of the peritoneum.

[1] Whether the alteration in the bony attachment is not rather apparent than real, there is good reason to doubt. The view of Fritsch, that the sacro-uterine ligaments have *no bony insertion*, but are limited posteriorly by the peritoneum is, to my mind, an error. It is possible that some of the muscular fibres inclosed in the folds of Douglas have no connection with the sacrum, but I do not think that it can be verified that all terminate within the peritoneum. The uterus, during gestation, unquestionably drags the folds of Douglas upward, but I doubt if the muscular layer beneath them (the true contractile elements of these folds) have their bony attachments displaced. It would seem to be in direct violation of all known laws pertaining to muscles.

[2] This view is not fully in accord with some late researches, which seem to demonstrate that the pelvic peritoneum is more readily separated from the pelvic wall than was formerly supposed. In gastro-elytrotomy, the fetus is removed beneath the peritoneum. During gestation, it also exhibits marked alterations from the normal standard, in respect to its pelvic attachments. Why, therefore, cannot blood extravasated beneath it be justly believed to lift the peritoneum, in those localities where that membrane can be separated from its bony relations in the virgin, without any marked difficulty?

The Pelvic Cellular Tissue.—If a plane be imagined as passing from the central point of the inner surface of the pubes to the point of junction of the third and fourth bones of the sacrum, the entire pelvic peritoneum (save a part of Douglas' pouch), would lie entirely above it. Beneath this imaginary plane, in the intervals which exist between the various viscera of the pelvis, ramify the blood-vessels, lymphatics, and nerves, and a collection of fibrous and fibro-elastic elements and muscular tissue, which compose the so-called cellular tissue of the pelvis. This imaginary plane may be said to divide the pelvic cavity into two spaces, the *peritoneal* and the *subperitoneal*. It is within the latter space that the structure which we are now to consider is exclusively confined. The proportions of the fibrous, fibro-cellular, connective, and muscular elements, which enter into the formation of the pelvic cellular tissue, vary; one or more predominating according to the *physiological functions* which it is destined to perform in any special locality. The functions of this tissue depend somewhat upon its situation. When it is placed in intimate relation to blood-vessels it acts, by its excess of muscular elements, as an important factor in the erecto-plexiform character of the venous system of the pelvis. When it is employed as a means of attachment for organs, it assumes more of a ligamentous structure, and becomes an important agent in preserving the mutual relations of the organs which it serves to connect, and also acts as a factor in preventing their displacement from their normal attitude. Some portions of this tissue act as lines of traction upon different portions of the uterus; some tend to keep the walls of the vagina in contact, since that tube is drawn both backward toward the sacrum, and laterally toward the sides of the pelvis; some maintain the Fallopian tubes in their proper relation to the ovaries; and some bind the ureters to the various structures with which they bear intimate relation.

This cellular tissue forms an integral part of the uterine

[1] The *recto-vesical fascia* must be considered as marking the lower limit of this space. This fascia lies immediately above, and in close contact with the levator ani muscle. A prolongation of the *obturator fascia* covers the inferior surface of this muscle. A diagram on a subsequent page (Fig. 22) will tend to make this arrangement clear to the intelligence of the reader.

system. It surrounds and constitutes a frame-work for the plexus of vessels which ramify within and upon the muscular structure of the uterus and vagina. It is continuous at its periphery with the sub-peritoneal cellular tissue of the walls of the abdomen. Those fibres, which may be considered as a "pubo-sacral process" of the entire mass, seem to be intimately associated with the chain of blood-vessels (chiefly venous) which begin at the bulbs of the vagina in front, and continue as plexuses upon the bladder, urethra, vagina, and uterus; while the muscular elements are continued still farther back in the sacro-uterine ligaments. Another set of fibres, which may be classed as the "utero-iliac process" of the entire mass, passes from the uterus along the vessels of the broad ligament to the sides of the pelvis, in a region somewhat anterior to the sacro-iliac synchondrosis, except during pregnancy.[1] It is thus easy to understand why the anterior and posterior layers of the broad ligament of the uterus are separated by this process of the pelvic cellular tissue; since its muscular fibres are sufficiently developed to afford that reflection of the peritoneal lining of the pelvis a power of resistance in excess of the simple elasticity of the peritoneum itself, in addition to the control exerted by them over the blood-supply of the womb. The bearings of the broad ligaments and also of the sacro-uterine ligaments upon uterine displacement have been discussed at some length in preceding pages.

Savage describes the cellular tissue of the pelvis as having been apparently thrust between the rectum and the bladder beneath the pelvic peritoneum, and as having fixed itself there by cellular attachments to every available part of the pelvic cavity. He makes use of the following clinical deduction: "The independence of this uterine collocation is exemplified in certain forms of sub-peritoneal abscess of long standing, which leave unaffected from first to last the rectum, ureters, and bladder."

The relations of the rectum and the upper portion of the urethra to this tissue are peculiar; each being inclosed in a complete investment or sheath which can be easily stripped off. In the median line, the cellular tissue between the peritoneal covering of the fundus uteri and the organ itself is so

---

[1] See pages which treat of the broad ligaments.

nearly wanting as to be of no importance from a clinical stand-point. Farre, in his researches, succeeded in removing this peritoneal coat only after prolonged maceration; and Goupil confirms this view in his statement that its presence can scarcely be determined between the peritoneum and the vagina and fundus uteri.

Viewed in the light of clinical experience, the experiments of König and Hewitt, as to the effects of the cellular tissue of the pelvis upon fluid or air injected beneath the peritoneum in different localities, have a special interest. It was found that when the injection was made near to the ovary or the Fallopian tube, the fluid passed slightly into the corresponding side of the pelvis by following the course of the psoas and iliacus muscles; when injected close to the upper and front portion of the cervix, it passed along the course of the round ligament of the uterus and subsequently into the pelvic cavity below the iliac region; when injected posteriorly to the cervix beneath the broad ligament, it passes into the posterior and lateral portion of the pelvic cavity, and afterward along the psoas muscle. These experiments enable us to appreciate the probable course of purulent collections within the pelvic cellular tissue. Pelvic abscesses may point or spontaneously open in one of the following regions: 1. In the iliac region. 2. In the suprapubic region, possibly reaching to the level of the umbilicus. 3. In the inguinal region, or thigh, following the psoas muscle. 4. In the perineum, at the side of the anus, having possibly escaped from the pelvis through the sciatic notch. 5. Into the vagina. 6. Into the cavity of the bladder. 7. Into the cavity of the rectum. 8. Into the cavity of the peritoneum.

The excessive development of cellular tissue between the layers of the broad ligaments renders this locality a frequent seat of these abscesses. They may sometimes be felt distinctly through the walls of the vagina and, if artificial evacuation be attempted, it is well to select this channel for that purpose. Such procedures are, however, not without danger. The remarkable size and anastomoses of veins within the pelvis point to one great risk, viz., the entrance of air into these channels. A case has been reported where sudden death followed an attempt to evacuate a collection of fluid within the region of the broad ligament—probably due to this cause alone. The subject of abscess of the pelvis has been considered at a greater length than space will here permit of, in the last edition of Thomas' standard work.

During gestation, this cellular tissue undergoes an enormous hypertrophy to fill the space left vacant by the ascent of the uterus and its broad ligaments. After parturition, the excess of this tissue is slowly absorbed, and the uterus and its broad ligaments gradually tend to resume their former relations to the other pelvic structures.

Points of special interest pertaining to the pelvic cellular tissue. As a summary of the previous pages, the following deductions may be drawn:

The pelvic cellular tissue lies underneath the peritoneum, padding the interstices between the organs and muscles of the pelvis, embracing the cervix uteri, and spreading from that point between the layers of the broad ligaments to reach the pelvic wall.

Our present knowledge of its disposition has been rendered more accurate in late years by the following methods of research: 1. By the careful examination of sections of frozen bodies and spirit-hardened pelves. 2. By injections of air beneath the peritoneum at various points, and afterward tracing its ramifications. 3. By the injection of water. 4. By the injection of plaster of Paris.

The first method affords a proper conception of the precise position of this tissue, its amount, and the area of its distribution in special topographical regions. Hart mentions the following sections as of value for this purpose: antero-posterior median (sagittal); antero-posterior lateral (lateral sagittal), intersecting the line of junction of the uterus with the broad ligaments; horizontal, at various levels; and, finally, the transverse section through the central point of the pelvis (coronal section).

The second method, assisted by the third and fourth, reveals the attachments of the cellular tissue to the pelvic peritoneum, and the tracts which normally exist between its various prolongations along which pus might burrow. In this way, our knowledge of the results of inflammatory attacks within the pelvis is placed upon a more positive footing, and our discrimination of the same made more precise.

The *retro-pubic fat* has been discussed in connection with the bladder and urethra. It is revealed by the sagittal section. It is triangular in shape. Its relation to the pubes is modified by the posture of the woman, especially in the genu-pectoral, in which it rises partly above the symphysis.

The *recto-vaginal process* extends between these tubes as low as the level of the pelvic floor. It is comparatively destitute of fat. It allows of the ever-varying degrees of distensibility of these tubes. It is best studied in the sagittal action of the pelvis.

The *vagino-vesical process* exists between the upper part of the anterior vaginal wall and the posterior surface of the bladder. Note that no such deposit of connective tissue exists between these organs in the lower two-thirds of the vagina. The small amount of tissue comprised in this process allows of a close approximation of the pelvic peritoneum to the upper portion of the anterior vaginal wall, when the bladder is empty—a point possessing surgical value.

The *rectum is separated from the sacrum* by a small amount of connective tissue, as seen in all sagittal sections.

Lateral sagittal sections show (1) a marked diminution of the retro-pubic fat; (2) that the laminæ of the broad ligaments are separated by a process of cellular tissue, studded with large vascular trunks, which steadily diminishes as the pelvic wall is approached; (3) that the base of the broad ligament touches the fornix vaginæ, except during gestation; (4) that the ovary bears an intimate relation with the anterior lamina of the broad ligament, while the posterior lamina is separated from it; (5) that the section of the round ligament of the uterus lies above that of the ovary, and anterior to it and the Fallopian tube.

Horizontal sections of the pelvis show (1) the retro-pubic fat; (2) the tissue of the ischio-rectal fossa; (3) the "parametric tissue" of Virchow and Spiegelberg; (4) abundant blood-vessels and lymphatics, imbedded in the parametric tissue, surrounding the lower part of the uterus and the upper part of the vagina.

Coronal sections reveal the relations of the pelvic cellular tissue and the ischio-rectal fossæ. The iliac, recto-vesical, obturator, and levator ani fasciæ, are also brought to view in their relations to the structures which form the female perineum. The respective outlines of the peritoneal, subperitoneal, and subcutaneous cavities are also rendered apparent. A diagram which I give in a subsequent page will enable the reader to grasp the somewhat intricate relations of these parts.

König's and Hewitt's researches, by means of subperitoneal injections, have been referred to, and need not be repeated. Bandl has also added to our knowledge, by similar experimentation. It would appear that fluid injected into the cellular tissue of the upper part of the broad ligament penetrates but to a slight extent into the cavity of the true pelvis, seeking rather a higher level in the neighborhood of the anterior wall of the abdomen. If injected, however, near its base, the peritoneum around the front of the cervix became raised, and the fluid followed the round ligament to the inguinal ring, and then passes into the iliac fossa; if indroduced posteriorly at the same level, the region of Douglas' space became first infiltrated, and then the region of the psoas muscle.

THE OVARIES.—Having discussed the reflections of the pelvic peritoneum, we are now capable of more accurately understanding the topographical anatomy of the ovaries. These bodies are best studied in the young virgin, because their surface is then smooth and their outlines more distinct than in the adult, when their surfaces are scarred by the rupture of the Graafian follicles. The ovaries are extremely movable, and their situation may be greatly modified by the relative position of the neighboring organs—chiefly that of the uterus and rectum. The wide variations in description with may be found to exist in the works of the more prominent anatomists is to be attributed largely to this fact.

These organs are analogous to the testes of the male and are classed under the head of follicular glands. Their size and shape may be compared to that of a small almond. They are usually found between the layers of the broad ligament, lying in the posterior pelvic space (see Figs. 7, 10, 19 and 20.) The inner extremity of each ovary lies about one inch from the lateral margin of the uterus and is connected to the cornu of that organ by a band of fibrous and muscular tissue, called the ovarian ligament. The outer extremity is connected to the fimbriated end of the Fallopian tube. The left ovary bears an intimate relation with the rectum. Each ovary is free on its two sides and also on its posterior border; but its anterior border is united to the anterior layer of the broad ligament, and at this spot the point of entrance of the vessels of the organ—the hilus—may be perceived. The measurements of the young virgin ovary are greater than at any other period of life, save during the first six weeks of utero-gestation, when the breadth and thickness of the organ are not altered, but its length is markedly increased. After parturition, the ovaries never again regain their former dimensions, unless from disease.

The two borders of the ovary are usually stated to be convex, but I have found the opinion of some of the later observers, viz., that the anterior border is practically straight, to be correct. This straight border is the attached one; and all other parts of the organ are freely movable. The investigations of Waldeyer have changed the prevalent opinion regarding the reflection of the peritoneum over the ovary; since it now seems to be well decided that the character of the epithelium changes as soon as the peritoneum is apparently reflected upon the ovary, thus transforming the investing coat of the organ into that of the mucous rather than serous type. It is now accepted, I think, as proven that the surface of the ovary must be considered as structurally continuous with the lining membrane of the Fallopian tube rather than with the peritoneum.

Most of the drawings of the ovary and its relative position to that of the uterus and the Fallopian tubes do not agree with my own observations, although the latter have been too limited to form a reliable basis for any positive deductions. I have almost invariably found the left ovary to be the smaller of the two. The surfaces of both organs have been more often flattened than markedly convex. The long diameter of the ovary, upon both the right and left side, I have found to lie usually in the vertical but sometimes in the antero-posterior diameter of the pelvis, in the virgin; seldom in the transverse pelvic diameter as is usually stated in text-books.[1] The left ovary

---

[1] This view is in full accord with the researches of His. Tait, Doran, Thornton, and others have likewise confirmed the frequency of the vertical attitude of the ovary.

was found to lie in contact with the rectal wall in almost every case examined. The point of attachment of the ovarian ligament to the uterus was usually found to lie between and below the attachments of the round ligament and the Fallopian tube.

My own observations have been especially directed toward the determination of another point in which I believe that most anatomists are in error, viz., the relation of the Fallopian tube to the ovary. In most of the accepted drawings of these parts, the Fallopian tube is represented as practically straight for some distance, as it passes from the uterus toward the pelvic wall, and then to terminate in its fimbriæ which hang considerably below its level, but in relation to the outer extremity of the ovary whose long axis is represented transversely. Now I have found in most of the virgins examined that this is not the case. On the contrary, the Fallopian tube has assumed rather the condition of an incomplete surcingle to the ovary; the fimbriated extremity passing so far below and around the ovary as to lie immediately beneath its convex border. If this condition be proven to be the normal one, the necessity for the grasping of the ovary by the fimbriæ during the escape of the ovum would seem to be wanting, as gravity alone would act as an important factor in causing the ovum to fall into a proper place for its prompt transmission to the uterine cavity by the Fallopian tube.

I am aware that this conclusion has been also arrived at by Doran, Tait and Thorton, and that some confirmatory observations respecting it have been lately offered in the London Pathological Society proceedings. It seems to me to be a point in topographical anatomy which will well repay extensive investigation, and I am glad that my own observations have been confirmed by others. Upon the left side, I found that the Fallopian tube did not dip so far down into the pelvis on account of the rectum, but that in some instances it nevertheless bore the relation to the ovary which I have described.

Points of special interest pertaining to the Ovaries.—The relation of the Fallopian tube to the ovary, as given on a previous page, will help to explain some clinical points which present themselves in cases of suppuration and dilatation of the Fallopian tubes and in tubal pregnancies.

As the fimbriæ of the tube run upward to reach the ovary, a slight attack of peritonitis might glue them to that organ. If the tube should become distended after such an agglutination, as in cases reported by Tait, the dilated pouch would be sausage-shaped and firmly connected to the uterus by the ovarian ligament, while the ovary itself would be included between the tumor and the uterus.

Tubo-ovarian cysts, for the same anatomical reason, lie either external to or below the ovary, and not above it as would be the case if the old idea of the normal course of the tube were correct.

Fetation of the outer third of the tube, must, of necessity, lie below the ovary and not above it, and might therefore be easily confounded with ovarian pregnancy.

Ovulation seems to create menstrual changes in the mucous lining of the uterine body, but exactly what these changes are must be considered as unsettled. Williams supports the view that the mucous coat is entirely destroyed at the menstrual epoch and that it is regenerated from the muscular coat, by cell-proliferation. Kundrat and Engelmann maintain that only the superficial layer of the mucous membrane of the uterus is shed, as a result of fatty degeneration. Möricke denies that any of the mucous lining is removed, since microscopical examination of the substance, removed by a curette from uteri of menstruating women, fail to sustain the statements of the authors previously quoted.

The dominant influence of the ovary upon menstruation has been disputed by some late authors, among whom Tait stands foremost. It is known that Battey's operation does not totally arrest the process, even where both ovaries are removed. It is claimed by Tait that menstruation will always cease, however, when the Fallopian tubes are likewise excised. This author believes that these tubes play an important part in the menstrual function which has hitherto been unsuspected.

The medico-legal bearings of the "corpus luteum" are based largely upon the valuable essay of Dalton.

The Fallopian tube is kept in its proper relations to adjacent parts by the ovarian and infundibulo-pelvic ligaments. The former measures about 3 cm. and the latter 2 cm. in length. The infundibulo-pelvic ligament is simply that portion of the upper margin of the broad ligament which is not occupied by the Fallopian tube.

THE CIRCULATORY ORGANS OF THE PELVIS.—We have now considered the topographical relations of all of the pelvic viscera, their peritoneal reflections, and the pelvic cellular tis ue. In many places the vessels have been casually referred to, when any point arose which had a direct bearing upon them, but no special description of them has as yet been given. The blood-vessels are always a source of fear to the operating surgeon, and their course and relations cannot be given in too great detail; but as lack of space precludes a complete description, I will call attention to such points as are of special surgical or physiological value, and leave the remainder to those reliable works which are at every reader's command.

The female pelvis is remarkable for the number of its venous channels and their extensive anastomoses. The uterus, rectum, vagina, bladder, and urethra have venous plexuses which almost completely invest them, and which, by their anastomoses with each other and the veins of the perineum, form a continuous chain of veins from the sacrum posteriorly, to the bulbs of the vagina in front. The broad ligaments of the uterus also inclose extensive venous plexuses.

The veins of the pelvis are almost entirely destitute of valves; hence their communications are of surgical importance from the danger of excessive hemorrhage which they entail upon any wound of the pelvic organs or even of the structures which compose the pelvic floor.

Besides, the veins of the pelvis serve other important purposes than as mere channels for the transmission of blood. In order to appreciate this statement, it is necessary to divide the pelvic structures into three classes: the erectile, the erecto-turgescent, and the turgescent. The *erectile* structures include the clitoris and its two crura; the *erecto-turgescent* include the body of the uterus, the bulbs of the ovary, and the bulbs of the vagina; the *turgescent* comprise the urethra and the vagina.

In the erectile structures, the circulation is of the ordinary character, except when the special functions of the parts are called into play; then they become enormously engorged with blood, and assume the condition termed "erection." In the turgescent structures, the blood supply is normally excessive, and suffers little temporary alteration in any of the physiological processes. In the erecto-turgescent bodies, the normal turgescence is supplemented, at times, by the condition of temporary erection, thus acting in concert with the erectile bodies. Each of these three classes of structures has a distinctive peculiarity as regards the arrangement of its blood-vessels which is characteristic. The chief features of these peculiarities will be found summarized in the excellent work of Savage, and given in more detail in the larger treatises upon histology.

Between the layers of the broad ligaments of the uterus, the spermatic vessels reach the hilus of the ovary of either side, at which point the artery enters and the veins escape. The spermatic veins anastomose with the sub-ovarian plexus of veins, and these, again, with the veins of the uterus, thus forming one continuous chain of large venous channels for the entire length of the broad ligament. The rupture of the sub-ovarian plexus is one of the most frequent causes of pelvic hematocele.

The uterine artery and veins lie in close relation with the folds of the broad ligament near to its base, and also with the vagina at its upper portion, and are in free communication with the vessels of the lower part of that tube. Along the sides of the uterus, between the folds of the broad ligaments, is found an enormous collection of veins, intermingled with spiral branches of the uterine artery, whose ramifications invest the entire organ. In a corroded preparation, as depicted in Savage's work, this arrangement of the vessels of the body of the uterus (chiefly venous) is beautifully shown. This collection of veins has a free anastomosis with the sub-ovarian plexus.

The uterine artery and a circular branch, which is sometimes present, might cause serious hemorrhage in operations upon the cervix. The vesical and vaginal branches are uncertain in their point of origin and distribution; hence they are liable to cause unexpected embarrassment in operations for vesico-vaginal fistula. The anastomosis between the spermatic and uterine arteries is of surgical importance in all procedures for the removal of the uterus, since a ligature applied to the cervix or to the vagina just below its uterine attachment would fail to control the hemorrhage from the spermatic artery. In those cases where the uterine artery is abnormally small, the spermatic artery is proportionately enlarged. The uterine artery passes between the layers of the broad ligament, at first near to its bony attachments; it then passes through the cellular tissue between the uterus and the pelvic wall in the base of the broad ligament; finally, it ascends between the layers of the broad ligaments, close to the lateral border of the uterus. It sometimes surrounds the ureter, like a cord.

THE FEMALE PERINEUM.—An article which treats of the anatomy of the female pelvic organs is not complete without some mention of the relative situation of the important structures which form the pelvic floor. This portion of regional anatomy in the female is either omitted entirely or inadequately described by most authors on anatomy. In the last edition of Thomas' treatise, the meagreness of our knowledge upon this subject has been referred to at some length. I may be pardoned, therefore, if I quote from quite an exhaustive article[1] upon the anatomy of this region, an extract which seems to me to cover the main points that my own researches have led me to believe and publish. The article itself will, if consulted, supply many deficiencies which may occur to the reader.

A GENERAL SUMMARY OF THE ANATOMY OF THE PERINEUM.— It is an impossibility to represent, in any one drawing, an accurate conception of the parts which have been discussed in the preceding pages. A sagittal section of the pelvis fails to give a true conception of the iliac fascia and the two lamellæ into which it divides at the level of the obturator internus muscle; while, on the other hand, this section is the only one which can be used in a diagrammatic way to represent the layers of the perineum, so as to properly appreciate their relations to each other. Savage has attempted to utilize a section of the pelvic structures made in the bis-ischiatic diameter of the pelvis, for the purpose of showing the relations of parts in the ischio-rectal fossa, but it strikes me as calculated, if used alone, to confuse rather than enlighten the reader. After innumerable

---

"The Female Perineum," by A. L. Ranney, New York Med. Jour., July and August, 1882.

attempts to devise a sketch which will fulfil the desired object, I have concluded to fall back upon the simplest form of diagram, which will enable the reader to grasp some of the more important points, without attempting to aim at any similarity to the parts as actually presented by dissection.

It will be seen that the cut shows the perineum and ischio-rectal space as adjoining one another, and inclosed by a continuation of the same layer of integument (a–a). The superficial layer of the superficial perineal fascia (b) likewise extends

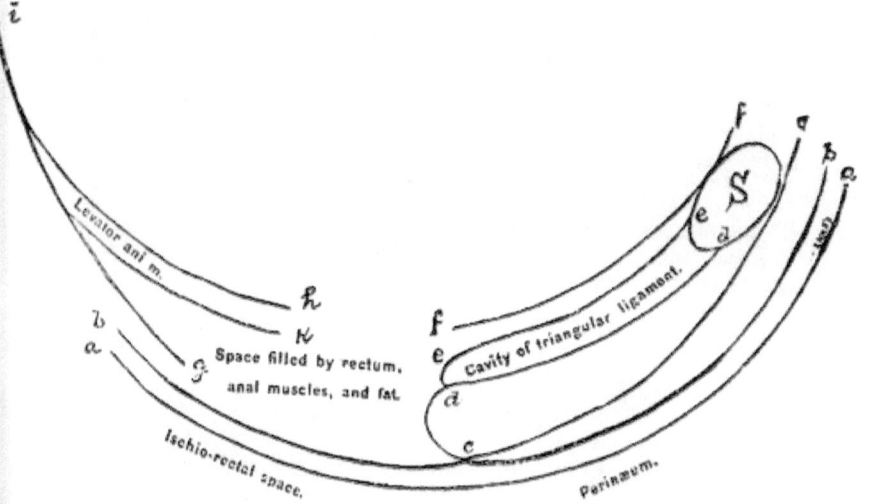

FIG. 21.—A Diagram designed to show the General Construction of the Female Perineum and Ischio-rectal Fossa. a, line of skin; b, line of superficial layer of superficial perineal fascia; c, deep layer of same; d, anterior layer of deep perineal fascia (triangular ligament of the perineum); e, posterior layer of same; f, iliac fascia covering pudic region; i, same fascia from the sides of the pelvis splitting into two lamellæ; g, obturator fascia; h, recto-vesical fascia; S, symphysis; k, fascia covering the levator ani muscle on its lower surface.

over both regions, but becomes adherent to the deep layer of the same fascia (c), at the line which divides these localities. Thus we have two layers which are common to both of these topographical regions. In the ischio-rectal space, we see that the *iliac fascia* sends two prolongations downwards—the *obturator fascia* (g) and the *recto-vesical fascia* (h), the former of which follows the pelvic wall, and incloses the obturator internus muscle, while the latter passes to the bladder and rectum, as its name indicates. The recto-vesical fascia bears an important relation with the levator-ani muscle, and is therefore of great importance here.

In the ischio-rectal space, there remains a large excess of room between the recto-vesical fascia and the pelvic floor, which is filled with the levator-ani muscle, the levator-ani fascia (k), and fatty tissue.

If we turn to the perineum proper, we shall perceive that several layers are depicted in the diagram, and that between these are left spaces which afford room for muscles, blood-vessels, nerves, etc. The dividing lines (b, c, d, e, f) depicted in the drawing are supposed to represent the different fasciæ. The various structures which are discovered between these fasciæ, as a dissection of the perineum is being made from without inward, can best be arranged in the form of a table, as follows:

Between the integument and the superficial layer of the superficial perineal fascia.
- Superficial hemorrhoidal vessels.
- Superficial hemorrhoidal nerves.
- Superficial perineal artery and nerve.
- Pudendal artery and nerve.

Between the deep layer of the superficial perineal fascia and the triangular ligament.
- Three pairs of muscles.
  - Bulbo-cavernosus.
  - Erector clitoridis.
  - Transversus perinei.
- Transverse perineal artery, vein, and nerve.
- Venous plexuses.
- Bulbs of the vagina or vestibule.
- Pudendal sacs.
- Dorsal artery and vein of clitoris.

Between the two layers of the deep perineal fascia (the cavity of the triangular ligament).
- Compressor urethræ muscles.
- Muscular fibres of the vagina attached to the rami.
- Pudic vessels and nerves.
- Urethra } perforating both layers.
- Vagina }
- Origin of four branches of pudic artery.
- Vulvo-vaginal glands.

Between the triangular ligament and the iliac fascia.
- Pubo-coccygeus muscle.
- Fibres of the levator ani muscle.

Between the obturator and recto-vesical fasciæ.
- External hemorrhoidal vessels and nerve.
- Levator ani muscle.
- Adipose tissue (in excess).
- Perineal process of the obturator fascia (covering the levator ani muscle), designated in the text as the "levator ani fascia."

of the Female Pelvic Organs. 103

The *ischio-rectal fossa* demands a special summary. As has been stated, it is impossible to properly represent its various component structures in a sagittal section of the pelvis. Perhaps the best view of this space can be obtained in a section of the pelvis made through the tuberosities of the ischia, although the relations of both the vagina and the rectum to

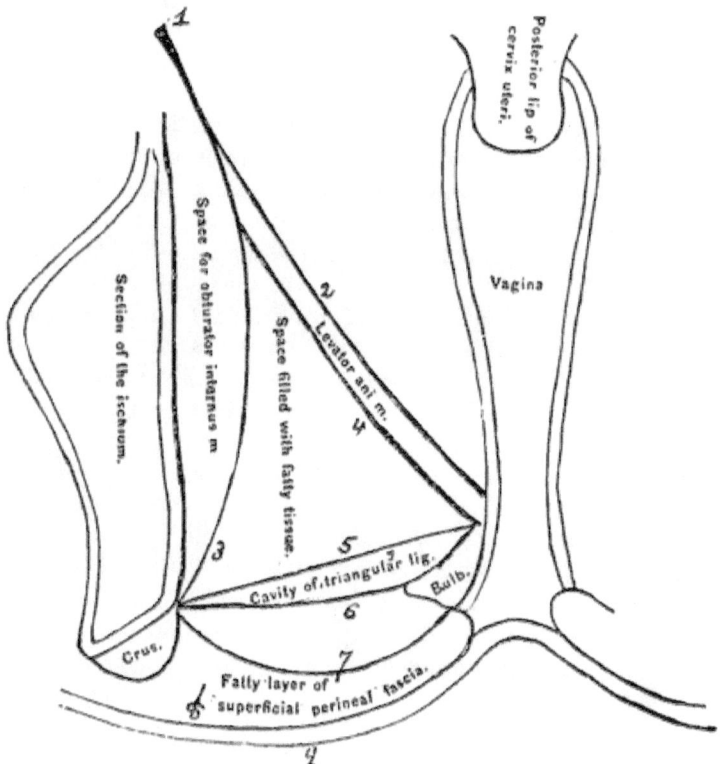

FIG. 22.—A diagram of a perpendicular pelvic section through the bis-ischiatic diameter. 1, iliac fascia; 2, recto-vesical fascia; 3, obturator fascia; 4, perineal process of obturator fascia, covering the under surface of the levator ani muscle; 5, deep layer of triangular ligament; 6, superficial layer of same; 7, deep layer of superficial perineal fascia; 8, superficial layer of same; 9, skin.

this space are not seen even then. The diagrammatic cut which is here introduced will, however, possibly assist the reader to grasp some of the most difficult points which such a section would help to make clear. It should be premised that this cut is purely diagrammatic, since the plane of such a section would cut the vagina in its transverse rather than its long diameter, and thus confuse the reader, if the section were properly represented. It is sufficiently accurate, however, for all practical purposes, and shows the general relation of parts even better than if made true to nature. It will be seen that the following fasciæ are met with in a dissection from above downward : 1st, the recto-vesical fascia; 2d, the dense fascia which lies beneath the levator ani muscle ; 3d, the two layers of the triangular ligament, situated below an extensive space filled with fat ; 4th, a muscular layer beneath it; 5th, the deep layer of the superficial perineal fascia and the muscles beneath it; 6th, the superficial layer of the same fascia with its deposit of fatty tissue ; 7th, the skin. The diagram shows, in addition, that the bulb of the vagina has a separate sheath of its own, as the crus of the clitoris has also, both of which are cut across. This fact alone would indicate that both were composed of erectile tissue.

From the diagrams and text of this article some deductions may be drawn which are applicable in many ways.

Removal of tissue about the vulva would sever the small terminal branch of the superficial perineal artery, which might require ligation. It might also loosen the attachments of the deep layer of the superficial perineal fascia, and thus disturb its functions; while inflammation might, moreover, extend upward from the pudendal sacs to the region of the inguinal canal.

The close proximity of the vaginal bulbs to the vulva, and the enormous collateral venous circulation which exists between the perineal veins and the intra-pelvic venous plexuses, might create alarming hemorrhage during an operation, or produce a granulating wound which it would be difficult to close with a firm cicatrix. This applies especially to the nymphæ.

The perineal body can be divided in the median line with scarcely any hemorrhage. Its vascularity increases as you pass outward toward the rami of the pubes and ischium.

Constriction of the vulvo-vaginal ring is produced chiefly by the looped fibres of the pubo-coccygeus muscle, and not by the bulbo-cavernosi muscles, as commonly taught. The term 'sphincter vaginæ' is improperly applied, therefore, to the latter muscles, because they are clearly analogous to the accelerator urinæ of the male, and also because such a function is not sustained by anatomical research. These latter muscles probably compress the bulbs and help to force an excess of blood into the clitoris.

The fatty tissue found in the superficial layer of the superficial perineal fascia may undergo such an enormous increase as to constitute irregular tumors of the pudendum.

No vessel of the perineum, if divided, requires a ligature at both ends except the internal pudic artery.

Small wounds of the vulva, vaginal orifice, or vaginal walls, especially during the pregnant state, may produce death from venous hemorrhage.

The dangers of incisions within the perineum increase as they approach the outer limits of that region, since the pudic vessels run close to the rami of the pubes and ischia, and their branches have increasing calibre as you pass from the median line of the body outward.

The superficial perineal artery differs in its course, size, and relations from the corresponding vessel of the male. The transverse perineal artery does not usually supply the vulvo-vaginal glands, as is commonly stated.

The vulvo-vaginal glands seem to lie, in some instances, posterior to the deep perineal fascia, while the glands of Cowper, which are their analogue in the male, are situated between its two layers. In others they were found to be identical, as regards their situation. Cunningham believes that this is to be considered as normal.

The levator ani muscle cannot be perceived in a dissection made from the integumentary surface inward until the superficial layer of the superficial perineal fascia is cut away, and the excess of fat which fills the ischio-rectal fossa as well as the perineal prolongation of the obturator fascia are likewise removed. It is seldom injured, therefore, in surgical procedures in the region of the female perineum, although it may be insolved in superficial incisions posterior to the bis-ischiatic line.

Actions of the Perineal Muscles.—The *bulbo-cavernosus* which is analogous to the lateral half of the accelerator urinæ muscle of the male, unquestionably assists in compressing the bulb of the vagina, and it may also similarly affect the vulvo-vaginal gland of the corresponding side. It is possible that it assists, as in the male, in creating and maintaining an erection of the clitoris by creating compression of the dorsal vein of the clitoris, and by forcing blood into the cavernous structure of that organ from the bulb of the vagina.

The *transversus perinei* serves to steady the central point of the perineum, when acting with its fellow, in order that the other muscles which are attached in that vicinity may have a firm point from which to act. It seems reasonable also to attribute to this muscle, as was first suggested by Cruveilhier, the function of antagonizing the action of the levator ani, which tends to draw the anus upward and forward.

Besides contracting the anal integument, the *external sphincter of the anus*, by virtue of its attachment to the perineal body and the tip of the coccyx, assists the levator ani in giving support to the opening during the expulsive efforts of defecation. Its fibres are intermingled with those of the bulbo-cavernosus muscle.

The *levator ani*, as its name would indicate, tends to raise and support the rectum and vagina during expulsive acts. In the female, the pubo-coccygeal portion also acts as the spincter muscle of the vaginal canal, and, possibly, of the urethra also after the vagina is collapsed. It is the physiological antagonist of the diaphragm in its action upon the pelvic viscera, as it rises and falls in unison with it during forcible respiration. When the action of the abdominal muscles is excessive, it yields, and thus enables the pelvis to bear a greater force than a more resistant structure; and, on the remission of such an action, it restores the perineum to its original form.

The *coccygeus* muscle helps to restore the coccyx to its normal position after it has been pressed backward during parturition or defecation.

The *erector clitoridis*, by its insertion into the sheath of the crus, may possibly assist in maintaining the erection of the clitoris by compressing the corpus cavernosum. Its size would seem to be in excess of that required to simply maintain a steadiness of that small organ, which is so necessary to the proper performance of the functions of the penis.

BLOOD-VESSELS OF THE FEMALE PERINEUM. — The blood supply to the perineal structures is derived from the branches of the internal pudic artery. It will be well, therefore, to trace the course of this vessel before describing its branches.

The *internal pudic artery* arises from the anterior trunk of the internal iliac; escapes from the pelvic cavity through the great sacro-sciatic foramen; re-enters it, after passing around the spine of the ischium, by means of the lesser sacro-sciatic foramen, and then gives off its branches. It is accompanied throughout its entire course by its veins and nerve.[1] As its relations are of importance in performing operations upon the perineum, more detail seems to be demanded as to the course of the main trunk and the general distribution of the vessels which take their origin from it. As it crosses the spine of the ischium, the gluteus maximus muscle and the great sacro-sciatic ligament cover it. From that bony point it traverses the cavity of the pelvis, lying to the outer side of the ischio-rectal fossa and upon the internal obturator muscle, till it reaches the ramus of the pubes, along which it then ascends. It is enveloped, in company with its nerve and veins, *in a sheath formed of the obturator fascia and a falciform process of the great sacro-sciatic ligament* for that portion of its extent where it lies in relation with the internal obturator muscle, which can be defined by measuring about one inch and a half from the anterior margin of the tuberosity of the ischium. As it ascends along the ramus of the pubes it pierces the posterior layer of the triangular ligament of the perineum, then passes for a short distance in a canal between the two layers, and subsequently perforates the anterior layer of the same fascia near the symphysis before it gives off its terminal branches—the artery of the corpus cavernosum and the dorsal artery of the clitoris. The anterior margin of the pubes is a guide to this vessel for the greater portion of its course after it enters the lesser sacro-sciatic foramen. Incisions approaching the rami are therefore associated with great danger of hemorrhage. This artery in the female is much smaller than in the male.

The branches given off by this large vessel within the perineum include the following: 1st, the inferior hemorrhoidal; 2d, the superficial perineal or vulvar artery; 3d, the transverse perineal artery; 4th, the artery of the bulb; 5th, the artery of the corpus spongiosum; 6th, the dorsal artery of the clitoris.[2]

[1] Two large veins usually accompany it. Savage, Ford, and Ellis give this as the normal number.

[2] In exceptional cases the internal pudic artery gives off a *vaginal* and *uterine* branch.

Many of these branches have been discussed to some extent in those pages which treat of the structures with which they bear relation. As the descriptions of the analogous vessels of the male will answer in most instances for those of the female, I will simply enumerate such points pertaining to each of these six branches as will bear directly upon what has preceded.

The *inferior* or *external hemorrhoidal arteries* comprise two or three small vessels which are given off from the internal pudic as that vessel crosses anterior to the tuberosity of the ischium. They run across the ischio-rectal fossa, through the mass of fat which helps to fill that fossa so as to bring its level up to that of the perineum proper, and are distributed to the sphincter and levator ani muscles, and the skin and parts about the anus. They are the chief sources of hemorrhage from all superficial wounds about the anus or the ischio-rectal fossa. These vessels have veins which accompany them and empty into the pudic veins.

The *superficial perineal* or *vulvar artery* is given off in front of the preceding branches. It pierces the obturator fascia and the anterior layer of the triangular ligament (deep perineal fascia), and then crosses the transverse perineal muscle to reach the triangular space between the bulbo-cavernosus and the erector clitoridis muscles. It then passes forward through the deep layer of the superficial perineal fascia, in which respect it differs from the male vessel. It is distributed to the vulva, and sends branches also to the muscles which are situated between the deep layer of the perineal fascia and the anterior layer of the triangular ligament. It is a source of arterial hemorrhage in wounds of the vulva. In the male, it supplies the scrotum, thus supporting the analogy between the vulva and a lateral half of the scrotum.

The superficial perineal artery is accompanied by two veins, and anastomoses with the pudendal twigs derived from the superficial pudic branches of the femoral artery.

The *transverse perineal artery* is a smaller branch than the one just described. It pierces the deep layer of the triangular ligament to reach the transverse perineal muscle, along whose cutaneous surface it is distributed, as well as to neighboring parts between the anus and the bulbs of the vagina. Like the preceding artery, it is situated beneath the deep layer of the superficial perineal fascia. It may be a source of hemorrhage in laceration of the perineal body, or wounds of the perineum which are forward of the anus, and not in the median line. It occasionally sends a small branch to the vulvo-vaginal gland, and usually one to the bulb of the vagina.

The *artery of the bulb* is a vessel of considerable size, but short. It arises from the internal pudic artery between the layers of the triangular ligament of the perineum, whose anterior layer it subsequently pierces, and sends branches to the bulb of the vagina of the same side. It also sends twigs which supply the meatus urinarius. This vessel has a surgical importance in the male which is wanting in the female, as it is a source of dangerous hemorrhage in lithotomy, if wounded. The erectile tissue of the vaginal bulb is partly supplied by this vessel. The close proximity of the bulbs of the vagina to the labia minora renders their removal a source of a troublesome wound, which cicatrizes slowly."

The two terminal branches of the internal pudic—the *artery of the corpus cavernosum* and the *dorsal artery of the clitoris*—are the supplying vessels of the erectile tissue of the clitoris. The former (a small branch) supplies the crus; while the latter (a vessel of larger size) passes along the dorsum of that organ and is distributed chiefly to the covering of the clitoris and the fibrous sheath of the crus. A free communication exists between the branches of both vessels. Sometimes both of these arteries are given off after the pudic artery has perforated the anterior layer of the triangular ligament near the sub-pubic ligament. When the clitoris is amputated, the two dorsal arteries may require a ligature. The vessels of the corpora cavernosa can usually be controlled by simple pressure, as the trabeculæ favor coagulation of the escaping blood, and thus tend to occlude the wounded arteries.

The *veins of the perineum* have been discussed, to some extent, on a preceding page. It is important to remember, however, that the veins of the perineum are afforded a free anastomosis with the intra-pelvic venous plexuses, since no valves exist in these vessels to prevent the flow of the venous current in any direction. These extensive anastomoses explain the occurrence of fatal hemorrhage from wounds of the vulva and vagina, as well as the surgical conditions designated as hematocele and varices. Savage enumerates a long list of these cases, where the symptoms are to be explained purely on anatomical grounds.

THE NERVES OF THE FEMALE PERINEUM.—The branches of the internal pudic nerve and the inferior pudendal branch of the small sciatic nerve are distributed to the perineum of the female, and follow about the same course as in the opposite sex.

The *pudic nerve* arises from the lower part of the sacral plexus, and escapes from the pelvis through the great sacro-sciatic foramen. It re-enters the pelvis, after passing around the spine of the ischium, through the lesser sacro-sciatic foramen, and follows the course of the internal pudic artery and its veins, with which it bears a close relation from the point where those vessels reach the border of the ramus of the ischium. It is covered, as is also the artery, by the obturator fascia during its passage through the ischio-rectal fossa; and, subsequently, by the anterior layer of the deep perineal fascia, after reaching the level of the perineum. The branches which escape from it are successively as follows: The inferior hemorrhoidal, superficial perineal, deep perineal, muscular, and the dorsal nerve of the clitoris.

The *inferior hemorrhoidal nerve* is occasionally derived from the sacral plexus. It is given off by the pudic nerve in the majority of subjects, however, and accompanies the vessels of the same name. It crosses the ischio-rectal fossa and is distributed to the external sphincter muscle of the anus and the integument of that region. It communicates, by means of its branches, with the superficial perineal nerve and the inferior pudendal branch of the small sciatic nerve.

The *superficial perineal nerves* are generally two in number. They are called the anterior and posterior; since their areas of distribution differ. The former send filaments of distribution to the integument of the labia, and the mucous covering of the clitoris, while the posterior sends branches chiefly to the skin of the anal region. The anterior nerve sends a filament to the levator ani muscle.

The *deep perineal nerve* is the larger of the two terminal branches of the pudic nerve. It accompanies the superficial perineal vessels, and is distributed to the integument of the vulva, the bulbs of the vagina, the vulvo-vaginal gland, and the muscles which are situated immediately beneath the deep layer of the superficial perineal fascia.

The *muscular filaments* of the pudic nerve are distributed to the bulbo-cavernosus, erector clitoridis, and compressor urethræ muscles. They are derived chiefly from the deep perineal nerve.

The *dorsal nerve of the clitoris* accompanies the pudic artery between the two layers of the triangular ligament, pierces the suspensory ligament of the clitoris, and terminates upon the dorsum of that organ. It anastomoses with the inferior pudendal branch of the small sciatic nerve. Savage states that the nerve supply of the clitoris is greatly in excess of that of the glans penis.

The pudic nerve is stated to have been successfully divided for the relief of that distressing condition, termed vaginismus. If Burn's operation be performed, as described by most authors on gynecology, it is certain that the pudic nerve is not divided, as the incision would fail to reach the situation of that nerve. In fact, the nerve lies so close to the internal pudic artery, that an incision would have to be carried almost to the extreme limits of the perineal boundary to expose it. There is only one spot in the entire course of that vessel where a ligature can be safely applied to it—viz., just before the artery enters between the two layers of the deep perineal fascia—and I doubt if the nerve could be safely divided elsewhere. The criticism of Thomas as to the efficacy of this step, as practised by Sir Jas. Simpson, seems to me to be well sustained on anatomical grounds.

CLINICAL POINTS SUGGESTED BY THE ANATOMY OF THE FEMALE PERINEUM.—The bearings of anatomy upon the conditions of pudendal hemorrhage, pudendal hematoma, pudendal hernia, urethral prolapse, abscess of the vulva and ischio-rectal space, cysts, and abscess of the vulvo-vaginal glands, vaginal prolapse, rectal prolapse, vesical prolapse, and urethral caruncle, are of the greatest importance, as tending toward a better knowledge of these diseases and the methods of treatment indicated.

Pudendal Hemorrhage.—The bulbs of the vagina, and the extensive anastomosis of large veins in the region of the vulva, account for the occasional occurrence of severe hemorrhage from wounds of that region. The erectile venous system of the female (whose exterior part comprises the vaginal bulbs and also veins, which communicate both with it and each other in the perineum) is afforded a free communication with the veins which form plexuses upon the internal pelvic viscera; hence, any wound of the region of the labium may draw blood from the venous plexuses of the pelvis, the superficial epigastric veins, the transverse perineal veins, the hemorrhoidal veins, and from all the other tributaries to the internal pudic vein.

This subject was first brought to professional attention in 1830 by Simpson, who reported several cases of fatal hemorrhage from wounds of the vulva. It has been known to follow muscular efforts, without an apparent injury,[1] although incisions, blows, and punctures are the most common causes. It seems probable that those cases where the hemorrhage has been very severe have been associated with a rupture or wound of the bulbs of the vagina. These erectile structures lie in close relationship with the internal borders of the labia minora, and tend to explain the hemorrhage which sometimes follows excision of those parts, and the difficulty often experienced in establishing cicatrization afterwards. The free anastomosis between the veins of the pelvis and perineum would seem to indicate an easy method of extension of inflammatory processes from superficial parts to the pelvic structures.

Pudendal Hematoma.—This condition is sometimes called "pudendal hematocele;" but the term "hematocele" is now confined, by long usage, to effusions of blood into serous sacs, and is, therefore, in my opinion, a misnomer. The condition might be designated as a "thrombus," with more regard to accustomed usage, although the name employed here seems to me to be the best one, because "thrombosis" is a term more commonly applied to the coagulation of blood within a blood-vessel. The surgical condition under consideration consists of a mass of blood within the labium or the areolar tissue which lies in close relation to the wall of the vagina. As a consequence, a tumor is discovered which fluctuates at first, but which tends to become hard as the blood coagulates. If suppuration be established, fluctuation may be again developed, and a tendency "to point" may be detected later. It seems to be well proven that this escape of blood is more common in the pregnant

[1] Simpson reports a case which was caused by the straining efforts during defecation.

female[1] than in the non-pregnant and virgin state. It is often the direct result of parturition, the bulbs of the vagina having, in some way, become ruptured. It may demand incision and the evacuation of the blood-clots.

Pudendal Hernia.—In the labia majora, there exists, as in the scrotum of the male, two sacs (one in each) which may admit of a hernical protrusion of intestine or omentum. These sacs, called the "pudendal sacs," have been described elsewhere by the author.[2] They bear an analogy to the scrotal cavity, although the peritoneum is not carried downward to form a separate sac, as in the male, by an organ descending from the abdomen. It is true that a prolongation of peritoneum (the canal of Nuck) invests the round ligament of the uterus as far forward as the lower limit of the inguinal canal, but it does not line the pudendal sac. This peritoneal pouch is usually obliterated in adults, but it may become distended by fluid, intestine, omentum, the bladder, ovary, or uterus. Intestine and omentum are the more common structures, however, which suffer extreme hernial displacement. It is seldom that the limits of a hydrocele effusion extend in the female below the anterior opening of the inguinal canal. Barnes has lately written up hernial displacements of the ovary, in a masterly style (see bibliography attached to this article).

Pudendal hernia is to be differentiated from a hydrocele of the canal of Nuck, abscess and tumors of the vulva, hematoma, and surgical affections of the vulvo-vaginal glands. It is possible for an inguinal hernia to become strangulated in the female by muscular spasm, edema, etc. A condition of permanent irreducibility may be developed, without strangulation, by a deposit of fat in the coats, or a simple thickening or adhesion of the coats themselves, provided the hernia be allowed to remain long unreduced.

Urethral prolapse.—In the enfeebled and the aged, the mucous coat of the urethra and its underlying connective tissue is occasionally subject to eversion, thus causing a red, sensitive tumor, at the situation of the meatus urinarius, which shows a tendency to bleed when handled. This condition has

---

[1] The pregnant state tends toward an enormous increase in the blood supply of the sexual organs. This may tend to explain the especial liability to rupture of vessels during this period.

[2] The Female Perineum. N. Y. Med. Jour., July and August, 1882.

been described by Guersant as occasionally present in young girls, since he states that he has seen it in fifteen cases between the ages of two and twelve years. It should be differentiated from the conditions of urethral polypus, irritable caruncle of the meatus, venous angioma of the urethra (as described by Savage), and villous growths.

Abscess of the vulva.—The labia are subject to phlegmonous and suppurative inflammation as the result of direct injury, irritating vaginal discharges, vulvitis, and certain states of the blood which particularly tend toward the development of anthrax and furuncle. If suppuration be manifest, an early incision will relieve the dangers of fistulous openings into the rectum, the extension of the suppurative process to the ischio-rectal fossa, and constitutional infection, as the parts seem to resist all tendency to spontaneous evacuation.

Cyst and abscess of the vulvo-vaginal glands.—We owe much of our knowledge of the diseased conditions of these glands to Huguier. Their anatomy has been made a subject of special investigation by Bartholin, after whom the glands are commonly named, and also by Morgagni, Huguier, and Duverney; but I am inclined to regard their description as incorrect. They are analogous to Cowper's glands of the male, but differ occasionally from them in their situation. They seem to be placed posterior to, and in some cases between, the layers of the triangular ligament, while Cowper's glands lie invariably between its two layers.[1] This is possibly to be explained by the fact that the cavity of the triangular ligament of the male is more commodious than that of the female. These glands open into the vulva, just in front of the attachment of the hymen, by small ducts (whose canal will admit a hog's bristle), the length of which is usually estimated as varying from three-fifths to three-quarters of an inch. Now, if the mouths of these small ducts become occluded from any cause, the secretion of the gland continues to form and is unable to escape; hence a cystic tumor is created by compression and absorption of the substance of the gland itself. Under certain circumstances these glands take on a suppurative condition, resulting in

---

[1] I have given in detail the results of my dissections of these glands in my article upon the female perineum. They differ markedly from those of the authors quoted.

abscess. In the case of cystic distention, the tumor will be circumscribed, elastic, and movable; it will also be much less painful than an abscess of these glands. Either condition must be differentiated from phlegmon of the labium by its small size and circumscribed limits, and from a boil by its apparent depth from the surface. Cysts are generally painless to the touch, and may exist for years without causing marked inconvenience. The close relation of this gland to the *deep perineal* branch of the internal pudic *artery* renders the operation for its extirpation a bloody one.

Prolapse of the vagina.—This condition is very rare, except in those who have borne children. Meigs, however, reports a case which occurred in a child, only a few months old, as the result of a convulsive attack; and a case is also mentioned by Cooper of the same condition being developed at the age of seventeen years. In spite of such exceptional cases, it may be said that pregnancy, which causes the vaginal structures to become greatly hypertrophied, tends especially to produce it. Anything which would create a relaxed condition of the anterior wall of the vagina or the sacro-uterine ligaments, might allow of the protrusion of the vaginal tissues through its own orifice, but the relations of these structures would necessarily entail upon such a displacement a simultaneous alteration in the position of the bladder, intestine, and uterus from the attitude assumed in health. It is common, therefore, to find this condition included by authors under prolapse of the uterus, cystocele, and enterocele. Considerable stress has been laid by Thomas upon laceration of the perineal body as one of the most frequent causes of prolapse of the posterior wall of the vagina and the anterior wall of the rectum. This is unquestionably a prominent factor in such a displacement. A glance at the diagram published by myself (Fig. 11) will enable the reader to appreciate a theoretical function of the perineal body to which Hart and others have called attention, viz., to assist in deflecting the feces through the anal canal (which forms almost a right angle with that of the rectum). He urges the great frequency of rectal prolapse after laceration of this body as one of the proofs of this function. Regarding the questions whether the vagina can be prolapsed without displacement of other organs of the pelvis, or whether the mucous coat of that tube can be prolapsed without disturbing its other coats, authors of note differ. From a purely theoretical stand-point, the opinion of Savage, that "prolapse of the vagina alone, or prolapse of the vaginal mucous membrane alone, are two affections which, anatomically considered, would seem impossible," is in accord with my own convictions.

Prolapse of the Bladder.—The condition, called "cystocele," may be evidenced, in bad cases, by a tumor appearing between the labia majora. From what is known of the topography of the bladder, it must be evident that no downward displacement can occur without an extreme relaxation of the anterior wall of the vagina, since that is its normal support below. It is usually present in women who have suffered some alteration in the normal tonicity of the parts from child-birth. The consequences of this deformity are to produce a pouch, from which the urine cannot be expelled on account of the internal urethral orifice being situated above the level of the most dependent part. This induces *decomposition of the urea* of the residual urine into carbonate of ammonia, which is a powerful irritant; hence a cystitis soon follows, as indicated by pain, heat, an increased frequency of micturition with vesical tenesmus, and a peculiar scalding sensation during the act. It has been maintained by Scanzoni, and some other authors, that this condition of the bladder may in some instances be the cause of the vaginal prolapse, and thus a primary disease rather than a sequel of vaginal relaxation. The question of diagnosis of cystocele is easily made by the employment of a sound introduced within the bladder.

Prolapse of the rectum.—This condition may exist as a consequence of laceration of the perineal body, or independently of it. The former variety has already been spoken of under the discussion of prolapse of the vagina. The independent variety may be the result of a general lack of the tonicity in the rectal walls, and also of all conditions which tend to produce excessive straining efforts, such as hemorrhoids, fecal impaction, polypi, etc. The tumor may be extremely large, and is to be diagnosed from hemorrhoidal tumors or polypi by the fact that it involves the whole circumference of the gut. In weak children, this condition is not infrequent.

Urethral caruncle.—At the meatus urinarius, and sometimes

for some distance along the wall of the urethra, little vascular excrescences occasionally appear, which render the urethral canal extremely irritable and occasion great suffering. They are very vascular, and are rich in nerves (Reid); hence they bleed freely, and are exceedingly sensitive to the touch. Wedl, who has investigated their structure, states that very consist of hypertrophied papillæ, and that an excessive development of areolar tissue accompanies this hypertrophy. Savage classes them as "pseudo-angiomata"; and claims to have discovered cysts filled with mucus within them. Age or parturition seem to exert no influence upon their development. They are to be differentiated from urethral prolapse, and syphilitic growths. Their situation often creates intense suffering when micturition is performed. Thomas reports cases of suicide from the misery entailed by these small excrescences. For some unexplained reason, removal of these growths is often followed by a return of the condition.

BIBLIOGRAPHY.

Adams, I. A.—A New Operation for Uterine Displacements. Glasgow Med. Jour., June, 1882.

Alexander, W.—A New Method of Treating Inveterate and Troublesome Displacements of the Uterus. Med. Times and Gazette, April, 1882.

Aran, F. A.—Leçons clin. sur les Maladies de l'Utérus. Paris, 1858.

Aran, F. A.—Etudes Anatomiques et Anatomo-pathologiques sur la Statique de l'Utérus. Paris, 1858.

Arnold, F.—Anatomie. Band II.

Barnes, R.—(1) Diseases of Women. London, 1873. (2) Hernia of the Ovary. Am. Jour. Obstet., Jan., 1883.

Bandl, L.—(1) Ueber Ruptur der Gebärmutter und ihre Mechanik. Wien, 1875. (2) Die Krankheiten der Tuben n. s. w. Stuttgart, 1879. (3) Die Krankh. der Tuben, der Ligamenta, und des Beckenperitonäums. Billroth's Handbuch d. Frauenkrh., Stuttgart, 1879.

Bantock, G. G.—Treatment of the Ruptured Perineum, London, 1878.

Beck, J. R.—How do Spermatozoa enter the Uterus. Am. Jour. Obstet., Nov., 1874.

Bécquerel, L. A.—Traité clin. des Mal. de l'Utérus. Paris, 1860.

Beigel, H.—Die Krankheiten des Weiblichen Geschlechtes. Erlangen, 1874.

Blandin, Ph. Fred.—Traité d'Anatomie topographique. Paris, 1834.

Blum, A.—Des affections de l'urèthre chez la femme. Arch. Gén. de Méd., 1877. Vol. II.

Boerner, E.—Ueber den puerperalen Uterus. Graz, 1875.

Bourgery, C., et Jacob, N. H.—Traité complet de l'Homme. Paris, 1839.

Braune, Wilhelm.—(1) Plates of. Translated by Bellamy. Phila., 1878. (2) Die Lage des Uterus und Fötus am Ende der Schwangerschaft. Leipzig, 1872. (3) Topog.-anatomischer Atlas. Leipzig, 1875.

Breisky, A.—Die Krankheiten der Vagina. Billroth's Handbuch der Frauenkrankheiten. Bd. VII., 1879.

Brickell, D. W.—The Proper Treatment of Pelvic Effusions. Am. Jour. Med. Sciences, April, 1877.

Budin, P.—(1) Recherches sur

l'Hymen et l'Orifice Vaginale.
Progrès Médicale, 1875. (2) Des
Lesions traumatiques chez la
femme, etc. Paris, 1878.
  Campbell, H. F.—Pneumatic
Self-replacement of the Gravid and
Non-gravid Uterus. Am. Gynecol. Trans., Vol. I.
  Credé, C.—Beiträge zur Bestimmung der normalen Lage der gesunden Gebärmutter. Archiv f.
Gynäk., Band I., 1870.
  Chadwick, J. R.—The functions
of the Anal Sphincters, So-called,
and the Act of Defecation. Trans.
Am. Gynecol. Soc., Vol. II., 1877.
  Chiara, D.—La Evoluzione spontanea sorpresa in alto mediante la
congelazione. Milan, 1878.
  Chrobak, R.—Die Untersuchung
der Weiblichen Genitalien u. s.
w. Billroth's Handb. d. Frauenkrankheit. Bd. I., 1879.
  Claudius, M.—On the Position
of the Uterus. Med. Times and
Gazette, 1865.
  Claudius, M.—Ueber die Lage
des Uterus. Ztschr. f. rationelle
Med., 3 Reihe, Band XXIII.,
1865.
  Colles, A.—Treatise on Surgical
Anatomy, Dublin, 1811.
  Courty, A.—Traité pratique des
Mal. de l'Utérus. Paris, 1872.
  Cruveilhier, J.—Anatomy of
Human Body. (Pattison), N. Y.,
1844.
  Cunningham, D. J.—The Dissector's Guide. Edinburgh, 1880.
  Currier, A. F.—The Normal
Posture of the Adult Uterus. N.
Y. Med. Jour., Jan. 27th, 1883.
  Depaul, J. A. H.—(1) Bul. de
l'Acad. de Méd., 1854. (2) Clinique
Obstétricale. Paris, 1876.
  De Sinéty.— Manuel pratique
de Gynécologie. Paris, 1880.
  Doran, A.—Brit. Med. Jour.,
Dec. 10th, 1881.
  Duncan, J. M.—(1) Researches in
Obstetrics. (2) Papers on the Female Perineum. London, 1879.
(3) Remarks on Lacerations of Orifice of the Vagina. Obstet. Jour.
Gt. Brit. and Ire., Vol. IV., 1876-77.
  Ellis, G. V.—Arrangement of
Muscular Substance in the Urinary
and Generative Organs of the Human Body. Med. Chirg. Tran.,
Vol. XXXIX.
  Emmet, T. A.—Principles and
Practice of Gynecology, Phila.,
1879.
  Engelmann, G. J.—The Mucous
Membrane of the Uterus. Am.
Jour. Obstet., May, 1875.
  Esmarch, F.—Plate of Sagittal
Section of Pelvis and its Organs.
Billroth's Hand. d. Frauenkrh.
(Breisky).
  Farre, A.—Article on Ext. Genitals of Female. Cyclopedia of
Anatomy. Supplement, 1859.
  Fort, C. H.—Anat. Differences
between White and Negro Races.
Am. Jour. Obstet., Vol. X., 1877.
  Foster. F. P.—(1) Topographical
Anatomy of the Uterus and its
Surroundings. Am. Jour. Obstet.,
Vol. XIII., Jan., 1880. (2) Mechanical Action of Pessaries. Am.
Gynecol. Trans., 1881.
  Frankhaeuser. — Die Nerven
der Gebärmutter. Jena, 1867.
  Freund, M. B.—Berliner klinische Wochenschrift, 1869.
  Fritsch. H.—Plate of Pelvic Section. See Mayrhofer. Billroth's
Handb. f. Frauenkrh.
  Fürst, L. (1) Plate of Pelvic Section. See Winckel. Billroth's
Handb. d. Frauenkrh., Bd. IX.,
page 7. (2) Die Maass- und Neigungs-Verhältnisse des Beckens.
Leipsic, 1875.
  Garrigues, H. J.—(1) Obstet.
Treat. of Perineum. Am. Jour.
Obstet., Apr., 1880. (2) Gastro-Elytrotomy. N. Y. Med. Jour.,
Oct. and Nov., 1878. (3) Rest after
Delivery. Am. Jour. Obstet., Oct.,
1880. (4) Additional Remarks on
Gastro-Elytrotomy. Am. Jour.
Obstet., Jan., 1883.
  Goodell, W.—(1) Lessons in Gynecology, Phil., 1879. (2) A Critical Inquiry into the Management
of the Perineum during Labor.
Am. Jour. Med. Science, 1871.
  Gray, H.—Anatomy, Descriptive and Surgical. Phil., 1880.
  Guerin.—Sur la structure des
ligaments larges. Comptes Rendus, 1879.
  Guersant, M. P.—Révue de Thórapeutique.
  Goupil, E.—Clin. Méd. des Mal.
des Femmes. Paris, 1860.
  Gussenbauer, C.—Ueber das Gefäss-System der äusseren weiblichen Genitalien. Sitzungsberichte der Wissenschaften. Juliheft, 1869.

Guthrie, C. F.—Anat. and Dis. of Genito-Urinary Organs.
Guyon, F.—(1) Moulds of Uterine Cavity taken to show its shape. Etude sur la cavité de l'utérus à l'état de vacuité. Jour. d. physiol.
Hart, D. B.—(1) Study of Two Mesial Vertical Sections of the Female Pelvis, etc. Edinb. Med. Jour., Apr.. 1879. (2) Structural Anatomy of the Female Pelvic Floor. London, 1880. (3) Note on Naked-eye Anatomy of Female External Genitals. Edinb. Med. Jour., Sep., 1882. (4) Physics of Bladder and Rectum. Edinb. Med. Jour., Sept., 1882.
Hart, D. B. and Barbour, A. H.—Manual of Gynecology. Edinburgh, 1882.
Heitzmann, C.—Anatomie des Menschen. Wien, 1875.
Henle, J.—Handb. der Eingeweidelehre des Menschen. Braunschweig, 1866.
Henle, J.—(1) Handb. der Anatomie des Menschen. 2te Auflage. Braunschweig, 1873. (2) Anatom. Atlas. Braunschweig, 1874.
Herman, G. E.—Vaginal Anteversion Pessaries. Lon. Obstet. Jour., Vol. VIII.
Heschl.—Zeitschrift der Wiener Aerzte, 1862, Vol. VIII.
Hewitt, G.—Mechanical Study of Uterine Pathology. London, 1878.
Hicks, J. B.—Displacements of the Uterus by Distention of the Bladder. Obs. Jour. Gt. Brit. and Ire., Vol. IV., 1876-7.
His, W.—(1) Schultze's Archiv f. Mikroscop., 1865. (2) Die Entwickelung des Hühnchens im Ei. Leipsic, 1868.
Hodge, H. L.—(1) Principles and Practice of Obstetrics. Phil., 1866. (2) Diseases Peculiar to Women. Phil., 1868.
Huguier, P. C.—Mémoir sur les maladies des appareils sécréteurs des organes génitaux externes de la femme. Paris, 1880.
Hyatt, H. O.—Normal Anatomy of the Vulvo-vaginal Orifice. Am. Jour. Obstet., Vol. X., 1877.
Hyrtl, J.—Handb. der topographischen Anatomie. Wien, 5te Auflage.
Joseph, L.—Beitrag zur Aetiologie der Uterusflexionen auf Grund anatomischer Untersuchung und klinischer Beobachtung. Beit. zur Geburtsh. u. Gynäk., Bd. II., 1879.
Kelsey, C. B.—On the So-called Third Sphincter of the Rectum. N. Y. Med. Jour., March, 1881.
Kiwisch, F. R.—Chapters on Diseases of the Ovaries (Clay). London, 1860.
Klein, E.—Die Aeusseren Genitalien. Stricker's Handbuch.
Klob, J. M.—Path. Anatomy of Female Sexual Organs. N. Y., 1868.
Kobelt, G. L.—Die Männlichen und Weiblichen Wollust-Organe des Menschen. Freiburg, 1844.
Kohlrausch.—Anatomie und Physiologie der Beckenorgane. Leipzig, 1854.
Kölliker, A.—(1) Entwickelungsgeschichte. (2) Beitrag zur Kenntniss der glatten Muskeln. Zeitschrift für Wissen., Leipzig, 1848.
Krause.—Muller's Archiv, 1832.
Kundrat and Englemann.—Untersuchungen über die Uterusschleimhaut. Stricker's Jahrbuch, 1873.
Kussmaul.—Von dem Mangel u. s. w. der Gebärmutter. Würzburg, 1859.
Küstner, O.—Untersuchung über den Einfluss der Körperstellung auf die Lage des nicht graviden, besonders des puerperalen Uterus. Arch. f. Gynäk., Bd. XV., 1879.
Le Bec.—Contribution à l'étude des ligaments larges. Gaz. Hebd., 15 Avril, 1881.
Lee, C. C.—Diagnosis of Pelvic Hematocele. Am. Jour. Obstet., August, 1873.
Lee, R.—Anatomy of Nerves of Uterus. London, 1841.
LeGendre, E. Q.—(1) Anatomie homolographique, Paris, 1868. (2) De la Chute de l'Uterus. Paris, 1860.
Leopold, C. G.—Die Lymphgefässe des normalen nicht schwangeren Uterus. Arch. f. Gynäk., Bd. VI., 1873.
Litzmann, C. C. T.—(1) Die Formen des Beckens, Berlin, 1861. (2) Wagner's Handwörterb. der Physiologie.
Longet, F. A.—Traité de Physiologie. Paris, 1868.
Lott.—Zur Anatomie und Physiologie der Cervix Uteri. Erlangen, 1872.

Luschka, H.—Die Anatomie des Menschlichen Beckens. Tübingen, 1864.

Lusk, W. T.—Science and Art of Midwifery. N. Y., 1882.

Madden, T. M—Lacerations of the Perineum. Am. Jour. Obstet., May, 1872.

Martin, E.—Ueber die physiologische Lage und Gestalt der Gebärmutter im lebenden Weibe. Ztschr. f. Geburtsh. u. Frauenkh., Bd. I., 1876.

Mayne, R.—Article on the Perineum, Cyclop. of Anat., Vol. III., 1839-1847.

Meyer, H.—Lehrbuch der Anatomie, 1873

Möricke. — Die Uterusschleimhaut in den verschiedenen Altersperioden. u. s. w. Ztsch. f. Geburtshülfe und Gynäk., Bd. III., 1881.

Müller, J.—(1) Ueber die Organischen Nerven der erectilen männlichen Geschlechtsorgane. (2) Archiv f. Anatomie, 1837.

Nonat, A.—Traité pratique des Maldes l'Utérus. Paris, 1874.

Oesterreicher und Erdl.—Atlas.

Olshausen, R.—(1) Die Krankheiten der Ovarien. Billroth's Handb. d. Frauenkh., Vol. VI., Stuttgart, 1877. (2) Ueber Dammverletzung und Dammschutz. Volkmann's Klinische Vorträge, No. 44.

Panas, M.—Recherches cliniques sur la direction de l'Utérus chez la femme adulte. Arch. Gén. de Méd., Vol. I., 1869.

Pausch, A.—Anatomische Bemerkungen über Lage und Lageänderungen des Uterus. Archiv f. Anat., Phys., und Wissensch. Med., 1874.

Pawlick.—Ueber das Sondiren der Ureteren der Weiblichen Blase u. s. w. Centralbl. f. Gynäk., October 15th, 1881.

Pfannkuch, W.—Archiv für Gynäk., Vol. III., 1872.

Playfair, W. S.—Science and Practice of Midwifery, London, '76.

Polk, Wm. M.—(1) Landmarks in the Operation of Laparo-elytrotomy. N. Y. Med. Jour., May, 1882. (2) Observations upon the Anatomy of the Female Pelvis. N. Y. Med. Jour., December, 1882.

Rainey. — Structure and Use of the Lig. Rotundum Uteri. Lon. Phil. Trans., 1880.

Ranney, A. L.—(1) The Female Perineum. N. Y. Med. Jour., July and August, 1882. (2) Descriptive Anatomy of Genital Organs of Female. Annals of Anat. and Surgery, April, 1883.

Rein, M.—Article on the Fundamental Nerve Plexus of the Uterus. See Bul. de la Société de Biologie.

Richet, A.—Traité pratique d'Anatomie. Paris, 1860.

Rouget, C.—(1) Recherches sur les Organes Erectiles de la Femme. Jour. de la Physiol, 1858. (2) Traité d'Anatomie. Paris, 1874.

Savage, H.—Anatomy of the Female Pelvic Organs. N. Y., 1880.

Sappey, Ph. C.—Traité d'Anatomie Descriptive. Paris, 1860.

Scanzoni, F. W. De.—Traité pratique des Mal. des Organes Sexuels de la Femme. Paris, 1858.

Schröder, K.—(1) Article in Ziemssen's Encyclopedia of Med., Vol. X. N. Y., 1875. (2) Schwangerschaft, Geburt, und Wochenbett. Bonn, 1867. (3) Ueber de normale Lage und die Lageveränderungen der Gebärmutter. Archiv f. Gynäk., Bd. IX., 1875.

Schultze, B. S.—Sagittal Section of Pelvis. See cut in Winckel. Billroth's Handbuch d. Frauenkh., Stuttgart.

Schultze, B.S.—(1) Ueber Versionen und Flexionen u. s. w. Archiv f. Gynäk., Bd. IV., 1872. (2) Ueber Anteflexionen der Gebärmutter und die Normale Lage der Gebärmutter. Ibid., Bd. VIII., 1875. (3) Zur Kenntniss von der Lage der Eingeweide im Weiblichen Becken. Ibid., Bd. IX., 1876. (4) Zur Frage von der pathologischen Anteflexion der Gebärmutter. Ibid., Bd. IX., 1876. (5) Die Exacte Ermittelung der Lage des Uterus in der lebenden Frau. Centralbl. f. Gynäk., Bd. II., 1878.

Schütz, A.—Plate of a sagittal section of a frozen subject with uterine prolapse. Archiv f. Gynäk., Bd. XIII.

Skene, A. J. C.—(1) Diseases of the Bladder and Urethra in Women. N. Y., 1878. (2) Glands of Female Urethra. Am. Jour. Obstet., Vol. XIII.

Simpson, A. R.—Effects of Traction on the Uterus. Edin. Med. Jour., October, 1879.

Simpson, A. R., and Hart, D. B.

—Relations of Abdominal and Pelvic Organs. Edinburgh, 1881.

Sims, J. M.—Clinical Notes on Uterine Surgery. N. Y., 1871.

Smythe, A. G.—The Position of the Hymen in the Negro Race. Amer. Jour. Obstetrics, Vol. X., 1877.

Spiegelberg, O.—Lehrbuch der Geburtshülfe. Lahr, 1877.

Stricker, S.—Handb. der Lehre der Gewebe.

Studley, W. H. — Mechanical Treatment of Flexions of the Uterus. Am. Jour. Obstet., Vol. XII., 1879.

Tait, Lawson.—Dilatation and Suppuration of the Fallopian Tubes. Proceedgs. Pathol. Soc. of London. Brit. Med. Journ., Dec 10th, 1881.

Tarnier et Chantreuil. —Traité de l'Art des Accouchements. Paris.

Taylor, G. H.—Diseases of Women. 1871.

Taylor, I. E.—Amputation of the Cervix Uteri. New York, 1879.

Taylor, I. E.—Structure of the Cervix Uteri. Am. Jour. Med. Sciences, Vol. LXXX.

Thomas, T. G.—(1) Anatomy and Physiology of the Female Perineum. Am. Jour. Obstetrics, April, 1880. (2) Diseases of Women. Philadelphia, 1882.

Turner, W. — Introduction to Human Anatomy.

Turnipseed, E. B.—Anatomical Differences between Negro and White Races." Am. Jour. Obstet., Vol. X., 1877.

Tyrrell, F.—Lectures on Anatomy before the Royal College of Surgeons, 1839.

Wagner, R. — Handwörterbuch der Physiologie.

Wadeyer, W.—(1) Eierstock und Ei, 1870. (2) Eierstock und Nebeneierstock. (3) Cut of sagittal pelvic section. See Billroth's Handb. d. Frauenkrh., Vol. II. (4) Die Eierstocks-Kystome. Archiv f. Gynäk., Bd. I.

Wedl.—Pathological Anatomy.

Wells,Spencer. —Sagittal Section of Pelvis of a Frozen Subject. See Chrobak, p. 7. Billroth's Handb. d. Frauenkrh.

Wernich. — Die Erectionsfähigkeit des unteren Uterus-Abschnittes. Beit. zur Geburtsh. und Gynäk., Bd. I.

Williams, J.—(1) Physiological Changes in the Position of the Healthy Unimpregnated Uterus. The Lancet. Vol. II., 1873. (2) Structure of Mucous Membrane of Uterus. London Obstet. Jour., Vol. II.

Wing, C. E.—Points Regarding Diagnosis of Uterine Anteflexion and Anteversion. Boston Med. and Surgical Jour., Vol. CVI.

Winkel, F.—(1) Krankheiten des Wochenbettes. Berlin, 1869. (2, Billroth's Handb. der Frauenkrh., Vol. IX.

Van de Warker, E.—Normal Movements and Position of the Unimpregnated Uterus. Am. Jour. Obst., Vol. XI., also N. Y. Med. Jour. Vol. XXI.

Velpeau, A. L. M.—Traité d'Anatomie Chirurgicale. Paris, 1826.

www.ingramcontent.com/pod-product-compliance
Lightning Source LLC
Chambersburg PA
CBHW022141160426
43197CB00009B/1384